Nils-Ole Walliser

Geometry and type IIB string theory

Nils-Ole Walliser

Geometry and type IIB string theory

Model building in type IIB superstring theory and
F-theory compactifications

Südwestdeutscher Verlag für Hochschulschriften

Impressum / Imprint

Bibliografische Information der Deutschen Nationalbibliothek: Die Deutsche Nationalbibliothek verzeichnet diese Publikation in der Deutschen Nationalbibliografie; detaillierte bibliografische Daten sind im Internet über http://dnb.d-nb.de abrufbar.

Alle in diesem Buch genannten Marken und Produktnamen unterliegen warenzeichen-, marken- oder patentrechtlichem Schutz bzw. sind Warenzeichen oder eingetragene Warenzeichen der jeweiligen Inhaber. Die Wiedergabe von Marken, Produktnamen, Gebrauchsnamen, Handelsnamen, Warenbezeichnungen u.s.w. in diesem Werk berechtigt auch ohne besondere Kennzeichnung nicht zu der Annahme, dass solche Namen im Sinne der Warenzeichen- und Markenschutzgesetzgebung als frei zu betrachten wären und daher von jedermann benutzt werden dürften.

Bibliographic information published by the Deutsche Nationalbibliothek: The Deutsche Nationalbibliothek lists this publication in the Deutsche Nationalbibliografie; detailed bibliographic data are available in the Internet at http://dnb.d-nb.de.

Any brand names and product names mentioned in this book are subject to trademark, brand or patent protection and are trademarks or registered trademarks of their respective holders. The use of brand names, product names, common names, trade names, product descriptions etc. even without a particular marking in this works is in no way to be construed to mean that such names may be regarded as unrestricted in respect of trademark and brand protection legislation and could thus be used by anyone.

Coverbild / Cover image: www.ingimage.com

Verlag / Publisher:
Südwestdeutscher Verlag für Hochschulschriften
ist ein Imprint der / is a trademark of
AV Akademikerverlag GmbH & Co. KG
Heinrich-Böcking-Str. 6-8, 66121 Saarbrücken, Deutschland / Germany
Email: info@svh-verlag.de

Herstellung: siehe letzte Seite /
Printed at: see last page
ISBN: 978-3-8381-3201-3

Zugl. / Approved by: Wien, TU, Diss., 2011

Copyright © 2012 AV Akademikerverlag GmbH & Co. KG
Alle Rechte vorbehalten. / All rights reserved. Saarbrücken 2012

Contents

1 Introduction   **5**
    1.1 Overview 7
    1.2 Type IIB flux compactifications 9
       1.2.1 Type IIB supergravity 9
       1.2.2 Moduli stabilization 11
       1.2.3 Intersecting D7-branes 15
    1.3 F-theory 16
       1.3.1 Sen's limit 20

2 Aspects of toric geometry   **22**
    2.1 The fan 22
       2.1.1 On the singularity and compactness of a toric variety 25
    2.2 Line bundles and toric divisors 26
       2.2.1 The Picard and the divisor group 26
       2.2.2 Polytopes and divisors 28
    2.3 Batyrev's construction of CY hypersurfaces 30
       2.3.1 Reflexive polytopes 30
    2.4 The intersection ring 32

3 A new offspring of PALP   **34**
    3.1 General aspects of `mori.x` 34
    3.2 Options of `mori.x` 37
    3.3 Structure of the program 49

4 Four-modulus 'Swiss cheese' chiral models   **51**
    4.1 Large volume scenario 53
       4.1.1 General idea 53
       4.1.2 Incorporation of D7-brane stacks 56
    4.2 Freed-Witten anomaly 56
    4.3 Instanton zero-mode counting 58
       4.3.1 Neutral zero-modes 58
       4.3.2 Charged zero-modes 59
    4.4 First model 59

		4.4.1 The resolved $\mathbb{P}^4_{15,10,2,2,1}(30)$ geometry	60

        4.4.1   The resolved $\mathbb{P}^4_{15,10,2,2,1}(30)$ geometry 60
        4.4.2   Scenarios in the first model 64
        4.4.3   Moduli stabilization analysis 69
   4.5   Second model 71
        4.5.1   R1 resolution of $\mathbb{P}^4_{2,1,6,1,2}(12)/\mathbb{Z}_2 : 1\,0\,0\,0\,1$ geometry 71
        4.5.2   Scenarios in the second model 72
   4.6   Summary and outlook 74

5 Toric constructions of global F-theory GUTs   77
   5.1   Construction of global models 79
        5.1.1   Setup 79
        5.1.2   Base manifolds 81
        5.1.3   Elliptically fibered Calabi-Yau fourfolds 87
   5.2   Data analysis 90
        5.2.1   Base manifolds 90
        5.2.2   Fourfolds 91
        5.2.3   Examples 96
   5.3   Summary and outlook 104

6 Restrictions on infinite sequences of type IIB vacua   106
   6.1   Type IIB moduli stabilization 108
        6.1.1   Calabi-Yau geometry 108
        6.1.2   Flux vacua 109
   6.2   Series in D-limits 111
        6.2.1   The no-go theorem of Ashok and Douglas 111
        6.2.2   D-limits 112
        6.2.3   D-limits and F-theory 112
   6.3   Series in type IIB D-limits 113
        6.3.1   Series around a large complex structure point 113
        6.3.2   Series in decoupling limits 116
        6.3.3   Series approaching a conifold point 117
        6.3.4   The two-parameter model $\mathcal{M}_{(86,2)}$ 119
   6.4   D-limits and infinite flux series for F-theory on $K3 \times K3$ 121
        6.4.1   F-theory with $G_{(4)}$ flux on $K3 \times K3$ 121
        6.4.2   The $K3$ surface 122
        6.4.3   D-limits and $\mathcal{G}_\Sigma$ 124
        6.4.4   Infinite series and automorphisms of $H^2(K3, \mathbb{Z})$ 126
   6.5   The models of Ahlqvist et al. 128
   6.6   Summary and outlook 132

Acknowledgements   135

A Appendix to chapter 4 — 137
A.1 Definitions and rules for B-branes 137
A.1.1 D-brane charges 137
A.1.2 Orientifolding 138
A.1.3 K-theory construction of D7-branes 140
A.2 Third model 141
A.2.1 R2 resolution of the $\mathbb{P}^4_{2,1,6,1,2}(12)/\mathbb{Z}_2 : 0\,0\,1\,1\,0$ geometry 141
A.2.2 Scenarios in the third model 144
A.3 Fourth model: a matterless model 145
A.3.1 The resolved $\mathbb{P}^4_{1,1,3,1,3}(9)/\mathbb{Z}_3 : 0\,0\,2\,1\,0$ geometry 145
A.3.2 Moduli stabilization 147

B Appendix to chapter 5 — 148
B.1 Matter genera and Yukawa points 148

C Appendix to chapter 6 — 151
C.1 Expansions around LCS points 151
C.1.1 One-parameter models 151
C.1.2 Coefficients of metric $\mathcal{G}_z$ of the two-parameter model 153

Bibliography — 154

Abbreviations

BI	Bianchi identity
BPS	Bogomol'nyi-Prasad-Sommerfield (bound)
CS	Chern-Simons
CY	Calabi-Yau
CWS	combined weight system
DSZ	Dirac-Schwinger-Zwanziger (product)
eom	equation of motion
FI	Fayet-Iliopoulos (term)
FW	Freed-Witten (anomaly)
GVW	Gukov-Vafa-Witten (superpotential)
GKP	Giddings-Kachru-Polchinski
GUT	Grand Unified Theory
ISD	imaginary self-dual
KK	Kaluza-Klein (reduction)
LCS	large complex structure (point)
lhs	left hand side
LVS	large volume scenario
MSSM	minimal supersymmetric Standard Model
MW	Majorana-Weyl (spinor)
NS	Neveu-Schwarz
PALP	package for analyzing lattice polytopes
R	Ramond
rhs	right hand side
SR	Stanley-Reisner (ideal)
SYM	supersymmetric Yang-Mills
YM	Yang-Mills
vev	vacuum expectation value

Chapter 1

Introduction

String theory is the best candidate for a theory that describes gravity at high energies. This is a quantum theory of gravity: it marries aspects of quantum field theory and general relativity in a consistent way. In this context, gravity is mediated via the graviton, which is a hypothetical massless elementary particle of spin two. The graviton naturally arises in the spectrum of quantized closed strings.

Furthermore, string theory unifies particles and unifies interactions. It does so by a change of paradigm with respect to quantum field theory. The concept of point-particle is replaced by a new idea: the fundamental degrees of freedom are the vibration modes of a one-dimensional extended object, the string. In a similar way to the point-particle propagating in spacetime along a world-line, the string sweeps out a two-dimensional surface, the world-sheet. The dynamics of strings is completely determined by an action proportional to the area of the world-volume. Particles arise as quantized excitation modes of the vibrating string. Interactions between strings are described in perturbation theory by joinings and splittings of world-sheets. More precisely, scattering amplitudes are given by infinite sums of topologically distinct world-sheets. These can be thought as thickened Feynman diagrams, but with two relevant differences. First, the diagrams are now classified according to their topology: a diagram of genus g corresponds to *all* g-loop Feynman diagrams. Second, interacting world-sheets describe smooth surfaces. In quantum field theory, interactions take place at points, which are topological singularities of the diagrams; this fact accounts for UV divergences. But now, intersection points are replaced by locally space-like, *smeared-out* regions – the points where the world-sheet seems to split are not Lorentz-invariant – and these divergences are cured.

The power of string theory lies in the fact that quantum consistency conditions severely restrict the possible formulations of string theory. For instance, conformal anomaly cancellation fixes the number of spacetime dimensions; bosonic string theory predicts 26 dimensions. Unfortunately, excitations of bosonic strings can not describe spin-1/2 particles. Since any realistic theory of fundamental interactions must also describe fermionic degrees of freedom at the end of the day, we need to equip the theory with supersymmetry on the world-sheet. This not only induces supersymmetry on the now ten-dimensional target space and thus defines fermionic superpartner fields, but has also a second important consequence; the spectra of the supersymmetric string (superstring in brief) are free of the tachyonic excitations that plague the bosonic string spectrum. It turns out that supersymmetry is a necessary condition for the consistent formulation of string theory.

The quantization of superstrings yields five different formulations of the theory:

- **Type I** is a ten-dimensional $\mathcal{N} = 1$ supersymmetric theory that contains open strings, which carry gauge degrees of freedom at their endpoints. Anomaly cancellation conditions require the gauge group to be SO(32). Type I is compatible with the presence of D1-, D5- and D9-branes.

- Type II are closed string theories with $\mathcal{N} = 2$ supersymmetry and no gauge group degrees of freedom. Depending on the relative chirality of right- and left-moving excitation modes of the string, type II give rise to **type IIA** (non-chiral) and **type IIB** (chiral) string theory. Both theories contain gravitational supermultiplets (graviton and gravitinos). Type IIA and IIB allow for the presence of D0-, D2-, D4-, D6-, D8-branes and D(-1)-, D1-, D3-, D5-, D7-branes, respectively.

- **Heterotic** string theories describe closed strings, whose left-movers correspond to excitations of the 26-dimensional non-supersymmetric bosonic string, whereas right-movers correspond to the modes of the ten-dimensional $\mathcal{N} = 1$ supersymmetric string. The excessive 16 dimensions are compactified on a self-dual lattice in such a way that ten-dimensional supersymmetry is preserved. There are only two types of lattice that satisfy this condition; they provide the non-supersymmetric field with gauge degrees of freedom **SO(32)** and $\mathbf{E_8 \times E_8}$, respectively.

Ten-dimensional supergravities are the low-energy, effective (tree-level) field theories of superstrings. Indeed, it can be shown that the massless spectrum of any superstring theory is equivalent to the spectrum of a ten-dimensional supergravity theory. In ten dimensions, there exist two distinct $\mathcal{N} = 2$ supergravity theories: type IIA and type IIB. Clearly, type IIA/B string theory reduces to type IIA/B supergravity. On the other hand, heterotic and type I strings reduce to type I gauged $\mathcal{N} = 1$ supergravities. There are several type I supergravities depending on the choice of the gauge group; but only those with $E_8 \times E_8$ and $SO(32)$ can be realized in string theory.

Furthermore, the five distinct superstring theories are connected via a chain of dualities. This fact suggests that there might be an underlying structure that unifies them. Indeed, from a modern point of view, we interpret these theories as different perturbative limits of an eleven-dimensional one that provides a non-perturbative description of strings; this theory goes under the name of M-theory. The following observation further corroborates this picture. M-theory reduces to the unique eleven-dimensional supergravity theory; type IIA supergravity can be obtained from dimensional reduction of the eleven-dimensional one.

Despite of its appealing theoretical aspects, no experimental evidence in favor of string theory is known at the moment. Even worse, we can only guess at which energy scales possible stringy phenomena would take place. Indeed, the string scale l_s is a (the only) free parameter of the theory; it can be thought as the length of the string. Nevertheless, if we want string theory to be a theory of quantum gravity, it seems reasonable to require the characteristic energy scale not to be much smaller than the Planck scale: $\lambda_s^{-1} \lesssim M_{Pl} = 10^{18}$ GeV. Under this assumption, it is fair to assume that no direct evidence of purely stringy phenomena can be achieved with the help of low-energy particle physics experiments. How do we proceed then in our attempt to investigate whether string theory is a valid physical hypothesis or just a fascinating mathematical framework? Surely, we do not have the arrogance to find a definitive answer here; we rather hope the present work can contribute to it.

Throughout our work we will follow this guideline: if string theory is a valid framework for fundamental processes, then its low-energy predictions must be consistent with established results from particle physics. In particular, we should be able to derive the Standard Model, or supersymmetric extensions thereof. Setting up phenomenological viable string models is one of the main goals of *model building*.

String theory predicts a ten-dimensional spacetime but, until now, experiments have given no indication of extra spatial dimensions. A way to make sense of the six extra (spatial) dimensions is to make them compact. The basic idea is to derive a four-dimensional effective theory via a process similar to the Kaluza-Klein reduction. The reduction sensitively depends on the geometry of the compact space. It is phenomenologically desirable that the ten-dimensional supersymmetry (partially) survives the compactification process. It turns out, that this constrains the geometry of the internal space to be a complex Kähler manifold with Ricci-flat metric, i.e. a Calabi-Yau manifold. Unfortunately, there is a huge number of such manifolds; it is not even clear if this number is finite. So, we end up with a plethora of possible low-energy effective theories, even if we started from a unique ten-dimensional formulation of sting theory. There is no first principle reason why to prefer a string vacuum instead of another. Things get even worse if we allow the presence of fluxes supported by D-branes. These add further freedom in the choice of vacuum configurations. Furthermore, note that the compactification process introduces a new characteristic length scale (roughly speaking, 'the radius' of the internal space) in the theory; this is another free parameter we have to cope with in addition to the string length.

We will focus on type IIB flux compactifications. These string vacua are relatively well-understood, and algebraic geometry ensures good control of their geometry. Indeed, even if the presence of fluxes generally destroy the Calabi-Yau condition, in type IIB they only partially affect it: the internal geometry remains Calabi-Yau up to warp factors.

1.1 Overview

In the present chapter, we hope to equip the reader with the necessary information for understanding the title of the present work; we review type IIB string compactifications and their relation to F-theory.

In chapter 2, we give an essential introduction to toric geometry focusing on those aspects that are relevant for type IIB orientifolds and F-theory model building. In the modern approach, a toric variety is described in terms of homogeneous coordinates, exceptional sets and a group identification. We discuss how these data are encoded in terms of convex cones. We furthermore explain the construction of subvarieties of toric ambient spaces in terms of lattice polytopes.

Chapter 3 presents the core computational techniques, which we make use of in the subsequent two chapters. Here, we discuss a computer assisted procedure aimed at constructing non-singular CY threefolds starting from reflexive polytopes, which computes their intersection rings and Chern classes. We discuss the program `mori.x`, which is part of PALP (a package for analyzing lattice polytopes) [1,2]. The program performs crepant star triangulations of reflexive polytopes and determines the Mori cones of the resulting toric varieties. Earlier versions of this program have been used to compute part of the results presented in chapters 4 and 5. This section is a refined version of the preprint article arXiv:1106.4529 [math.AG] and chapter 7 of the official PALP documentation arXiv:1205.4147 [math.AG].

In chapter 4, we discuss a very efficient strategy aimed at stabilizing Kähler moduli: the large volume scenario (LVS) [3]. We present compact, four-modulus 'Swiss cheese' CY threefolds that accommodate the LVS. In this type of compact spaces, the overall volume is driven by a single four-cycle, whereas the other cycles contribute negatively to it. These CYs are constructed as hypersurfaces embedded in toric fourfolds. We attempt to realize MSSM-like configurations on magnetized D7-branes within the LVS; we pay special attention to the chirality problem pointed out by the authors in [4]. We extend their analysis by properly taking into account the Freed-Witten anomaly on non-spin cycles. These constraints turn out to be very restrictive on our models. This chapter is an updated version of the article **JHEP 0907 (2009) 074** arXiv:0811.4599 [hep-th].

In chapter 5, we focus on the construction of a large number of compact CY fourfolds that accommodate global F-theory GUT models. The fourfolds are obtained as elliptic fibrations over non-CY base manifolds. The latter are constructed as hypersurfaces in four-dimensional toric ambient spaces. With the help of toric techniques, we search for divisors capable of supporting F-theory GUTs. In particular, we check whether the base space is regular and contains del Pezzo divisors. We further test the existence of mathematical and physical decoupling limits for each model. In the end, we are left with about 4 000 fourfold geometries. We construct $SU(5)$ and $SU(10)$ GUT models on every del Pezzo divisor. Carrying out this procedure, we obtain more than 30 000 models. This chapter is a refined version of the article **JHEP 1103 (2011) 138** arXiv:1101.4908 [hep-th].

Chapter 6 presents new results on the study of the so-called string landscape concerning the existence of a vast number of metastable four-dimensional vacua. One part of the landscape that is accessible by accurate analytical and numerical methods is the complex structure moduli space of type IIB flux compactifications. Ashok and Douglas proved [5] that infinite sequences of type IIB vacua with imaginary self-dual flux can only occur in special degenerate points of the complex structure moduli space, the D-limits. We refine this no-go result. We study a class of one-parameter CYs and show that there is no infinite sequence of vacua accumulating at their D-limits. We corroborate the result with a numerical study of the sequences. This chapter is an updated version of the article **JHEP 1110 (2011) 091** arXiv:1108.1394 [hep-th].

1.2 Type IIB flux compactifications

In this section, we briefly review the field content of type IIB supergravity and the difficulties in the construction of four-dimensional Minkowski vacua. In particular, we address the stabilization of closed string moduli. In the end, we show the importance of orientifold planes in circumventing these problems. The interested reader may want to consult the review articles [6,7] for more details on flux compactifications. Further, for a very comprehensive exposition of type II orientifold constructions see [8].

1.2.1 Type IIB supergravity

Closed string theories are classified as type II. Their field content depends on the boundary conditions one requires on the left- and right-moving fermionic modes of the world-sheet, respectively. There are two possible choices: periodic boundary conditions, called Ramond (R) conditions, and anti-periodic ones, referred to as Neveu-Schwarz (NS) conditions. For closed strings, the physical states are constructed by tensoring left- and right-movers:

$$(\text{vector + MW spinor}) \otimes (\text{vector + MW spinor}) . \tag{1.1}$$

This gives four possible compositions of boundary conditions. The tensor product of left- and right-moving excitations with same boundary conditions, R⊗R and NS⊗NS, describes spacetime bosons, whereas fermions correspond to the mixed sectors R⊗NS and NS⊗R. Furthermore, two spinors can have either the opposite or the same chirality. In the first case, the resulting theory is type IIA superstring. This is a non-chiral theory with (1,1) local supersymmetry, i.e. its spectrum is symmetric under the exchange of left- and right-movers and hence preserves parity. On the other hand, if the left- and the right-moving spinors have the same chirality, we obtain type IIB superstring. This theory is chiral with (2,0) local supersymmetry and violates parity.

The massless spectra of type IIA and IIB superstrings form the multiplets of type IIA and IIB supergravity theories. It is often much less complicated to carry out calculations in the supergravity approximation rather than in the full-fledged string theory. Therefore, whenever possible, we choose to work in the low-energy effective theory. This is indeed a useful approach for many questions related to model building.

The NS-NS sector contains the ten-dimensional metric g_{MN}, the so-called graviton field. Furthermore, there are the antisymmetric Kalb-Ramond two-tensor $B_{(2)}$ (or simply B-field) and the dilaton ϕ, which is a scalar; these two fields also appear in the Dirac-Born-Infeld (DBI) action that describes the propagation of open string degrees of freedom. The exterior derivative

Sector	IIA	IIB
NS⊗NS	$g_{\mu\nu}$ $B_{(2)}$ ϕ	$g_{\mu\nu}$ $B_{(2)}$ ϕ
R⊗R	$C_{(1)}, C_{(3)}, \ldots, C_{(7)}$	$C_{(0)}, C_{(2)}, \ldots, C_{(8)}$
NS⊗R	Ψ_M λ	Ψ_M λ
R⊗NS	Ψ'_M λ'	Ψ'_M λ'

Table 1.1: Bosonic and fermionic massless field content of type IIA and type IIB superstring theory. Both theories have the graviton $g_{\mu\nu}$, the Kalb-Ramond two-form $B_{(2)}$ and the dilaton field ϕ. Their R-R sectors differ: type IIA and IIB contain gauge potential form fields $C_{(i)}$ of odd and even degree, respectively.

acting on the B-field gives rise to the three-form field-strength

$$H_{(3)} = dB_{(2)}, \qquad H_{MNR} = 3\partial_{[M} B_{NR]}. \tag{1.2}$$

The R-R sector contains the gauge potential forms that couple to Dp-branes via the Chern-Simons (CS) action. Type IIA theory has only gauge fields of odd degree ($C_{(1)}, C_{(3)}, \ldots, C_{(9)}$), whereas type IIB those of even degree ($C_{(0)}, C_{(2)}, \ldots, C_{(8)}$). For each gauge potential there is an associated field-strength $F_{(p+1)} = dC_{(p)}$. Electromagnetic duality implies a relation between $C_{(8-p)}$ and C_p that can be stated in terms of the field-strengths:

$$*F_{(10-p-1)} = F_{(p+1)}. \tag{1.3}$$

These constraints reduce the degrees of freedom of the gauge potentials by one-half. In particular, the middle-dimensional field-strength is self-dual:

$$*F_{(5)} = F_{(5)}. \tag{1.4}$$

The NS-R and R-NS sectors contain the fermionic fields. In each sector, there is a supersymmetric partner of the graviton and of the dilaton, which are called gravitino and dilatino respectively. Table 1.1 summarizes the field content of ten-dimensional type IIA and IIB supergravity.

The bosonic action of type IIB supergravity in the Einstein frame is[1]

$$S_{Bose}^{IIB} = \frac{1}{2\kappa^2} \int d^{10}x \sqrt{-g^E} \left[R - \frac{1}{2} \frac{\partial_M \tau \partial^M \bar{\tau}}{(\operatorname{Im}\tau)^2} - \frac{1}{2} \frac{|G_{(3)}|^2}{\operatorname{Im}\tau} - \frac{1}{2}|\tilde{F}_{(5)}^2| \right]$$
$$+ \frac{1}{8i\kappa^2} \int C_{(4)} \wedge G_{(3)} \wedge \bar{G}_{(3)}, \tag{1.5}$$

where we have defined new forms by mixing the R-R and NS-NS sectors:

$$\tau = C_{(0)} + ie^{-\phi}, \qquad G_{(3)} = F_{(3)} - \tau H_{(3)}, \qquad \tilde{F}_{(5)} = F_{(5)} - \frac{1}{2}C_{(2)} \wedge H_{(3)} + \frac{1}{2}B_{(2)} \wedge F_{(3)}. \tag{1.6}$$

g_{MN}^E is the Einstein metric and it is assumed to appear in all index contractions. This metric is related to that in the string frame via rescaling with the reciprocal of the square root of the

---
[1] $|F_{(p)}|^2 := \frac{1}{p!} F_{M_1 \ldots M_p} \bar{F}^{M_1 \ldots M_p}$.

string coupling (that is related to the vev of the dilaton field $g_s = e^{\phi_0}$):

$$g^E_{MN} = e^{-\frac{\phi}{2}} g_{MN} \,. \tag{1.7}$$

In what follows, we will suppress the upper index 'E' in order to avoid clutter in the notation. It should, however, be clear from the context which frame is appropriate.

The first summand of equation (1.5) is the well-known Einstein-Hilbert action with the ten-dimensional Ricci scalar R. The gravitational coupling in ten dimensions is given in terms of the string length:

$$2\kappa^2 = \frac{1}{2\pi}\left(4\pi^2\alpha'\right)^4 = \frac{\ell_s^8}{2\pi} \,. \tag{1.8}$$

The second summand of the action is the kinetic term of the axio-dilaton field τ. Then, the Maxwell terms of the generalized field-strengths $G_{(3)}$ and $\tilde{F}_{(5)}$ follow. The last part of the action is composed of wedge products and hence is independent on the metric; this is the CS term. Finally, note that expression (1.5) does not describe the dynamics of type IIB supergravity completely. Indeed, the self-duality condition on the five-form field-strength does not derive from the eom's and has hence to be added as a supplementary constraint. Taking into account the definition of $\tilde{F}_{(5)}$ in expression (1.6), equation (1.4) generalizes to

$$*\tilde{F}_{(5)} = \tilde{F}_{(5)} \,. \tag{1.9}$$

1.2.2 Moduli stabilization

In the low-energy limit, string theory gives rise to ten-dimensional supergravity compactified on a six-dimensional manifold, also called internal space. In this effective theory, we can think of the ten-dimensional spacetime as composed of copies of the internal space X attached at each point of the four-dimensional Minkowski spacetime; locally it takes the form: $\mathcal{M}^{10} = \mathbb{R}^{3,1} \times X$. Phenomenological models often require that part of the supersymmetry survives the compactification process. The preservation of $\mathcal{N} = 1, 2$ supersymmetry implies that there exists a globally covariant constant spinor on X. This conditions puts severe restrictions on the geometry of X: it can be shown [9] that, in absence of fluxes, the internal space needs to be complex Kähler with vanishing first Chern class. This kind of spaces are named after two mathematicians, Calabi and Yau, who respectively conjectured and proved the following statement. If X is a compact Kähler manifold with Kähler form J and vanishing first Chern class, then there exists a unique Kähler form $\hat{J}$ in the same cohomology class as J, whose corresponding Ricci form is zero. Or, to state it more simply, there is a unique Ricci-flat Kähler metric in each Kähler class.

Calabi-Yau manifolds come in families smoothly related to each other by deformation parameters called moduli. These parameters control shape and size of the CY manifolds. An important property of CY threefolds is that their geometrical moduli space is the product of two disjoint parts $\mathcal{M} = \mathcal{M}^{2,1}_{\text{CS}} \times \mathcal{M}^{1,1}_{\text{K}}$. The complex-structure moduli account for deformations of the shape. The Kähler moduli, instead, control the sizes of the threefold and of its subspaces. These moduli give rise to massless fields in the effective theory in four dimensions via a process similar to the Kaluza-Klein reduction. The dimension of the moduli space is determined by the Hodge structure of the CY threefold: $\dim \mathcal{M}_{\text{CS}}^{2,1} = h^{2,1}$ and $\dim \mathcal{M}_{\text{K}}^{1,1} = h^{1,1}$. A generic CY threefold comes with many moduli that, after KK reduction, lead to unwanted massless scalar fields in the four-dimensional theory. A possible way to overcome this problem is to introduce scalar potentials – induced by appropriate fluxes – that stabilize these fields at energies beyond the characteristic compactification scale. In this construction, the internal geometry is still (conformally) CY even after having turned on the fluxes; this property is crucial for controllability. The realization of this strategy is one of the main issues of type IIB flux compactification.

No-go theorem

For phenomenological reasons, we would like to find solutions of (1.5) from which four-dimensional effective actions can be constructed that preserve Poincaré symmetry. These solutions satisfy necessary conditions on the metric g_{MN} as well as on the form of the fluxes. In this context, the most general ansatz of the metric is

$$g_{MN} = \begin{pmatrix} e^{2A(y)}\eta_{\mu\nu} & 0 \\ 0 & e^{-2A(y)}\tilde{g}_{mn}(y) \end{pmatrix}. \tag{1.10}$$

Here, $\eta_{\mu\nu}$ is the four-dimensional Minkowski metric. The coordinate y parametrizes the internal manifold X that we assume to be a CY threefold with metric $\tilde{g}_{mn}(y)$. The warp factor $A(y)$ controls the relative sizes of different regions of the six-dimensional internal space. If fluxes are tuned off, it can be shown that $A(y) = 1$ must hold; in this case the four-dimensional spacetime reduces to the Minkowski space. In addition to ansatz (1.10), the axio-dilaton should depend only on the compactification manifold and $G_{(3)}$ should have only compact components and hence be supported on cycles of the internal manifold. It turns out that $G_{(3)}$ is an element of the third cohomology of X with integral values due to the flux quantization; furthermore, the self-dual five-form should fill the four-dimensional spacetime completely and extend in one compact direction:

$$\tau = \tau(y), \qquad G_{(3)} \in H^3(X, \mathbb{Z}), \qquad \tilde{F}_{(5)} = (1+*)\left[d\alpha(y) \wedge \underbrace{dx^0 \wedge \ldots \wedge dx^3}_{d\text{Vol}_4}\right], \tag{1.11}$$

where $d\text{Vol}_4$ is the volume-form of the Minkowski spacetime and $\alpha \in H^0(X)$ is a function of y. The Hodge star in the definition ensures the self-duality of the five-form.

We obtain Einstein equations by varying action (1.5) with respect to the metric. Then, we solve the ten-dimensional Ricci scalar in terms of the energy momentum tensor. We contract with $\eta^{\mu\nu}$ and obtain

$$\tilde{\nabla}^2 e^{4A} = e^{2A}\frac{1}{2}\frac{|G_{(3)}|^2}{\operatorname{Im}\tau} + e^{-6A}\left(|\partial\alpha|^2 + |\partial e^{4A}|^2\right), \tag{1.12}$$

where $\tilde{\nabla}$ is the covariant derivative with respect to $\tilde{g}_{mn}$. On a compact manifold, the lhs integrates to zero being a total derivative, whereas the rhs is composed of three non-negative terms. Hence, the equality holds only if each of these terms become zero. This implies that the solutions must have $G_{(3)}$ equal to zero and A constant. In other words, only trivial warped compactifications allow for solutions that preserve four-dimensional Poincaré symmetry. This result is an instance of the no-go theorem by Maldacena and Nuñez [10]. They considered the most general metric ansatz that preserves maximal symmetry in four dimensions. Their ansatz can be obtained from (1.10) by substituting the Minkowski metric with a generic four-dimensional metric $\tilde{g}_{\mu\nu}(x)$. They showed that the presence of fluxes is compatible with four-dimensional anti-de Sitter solutions, but inconsistent with Minkowski or de Sitter.

GKP's evasion strategy

The no-go theorem proves that the four-dimensional geometry cannot be Minkowski if we only include fluxes in type IIB compactifications. This is a very general statement; in particular, supersymmetry of the vacuum solutions is not assumed. At the same time, this result suggests a possible solution to overcome this bottleneck. The addition of a negative term on the rhs of (1.12) allows to consider non-trivial $G_{(3)}$ and warp factors without violating Poincaré symmetry. This strategy was first developed by Giddings, Kachru and Polchinski (GKP) [11].

String theory contains local objects, like D-branes and O-planes, whose contribution of non-perturbative nature can be added to the action (1.5): $S = S_{Bose}^{IIB} + S_{loc}$. With this inclusion, the eom (1.12) becomes

$$\tilde{\nabla}^2 e^{4A} = e^{2A} \frac{1}{2} \frac{|G_{(3)}|^2}{\operatorname{Im} \tau} + e^{-6A} \left(|\partial\alpha|^2 + |\partial e^{4A}|^2 \right) + \kappa^2 e^{2A} \underbrace{\left(T_m^m - T_\mu^\mu \right)_{loc}}_{T_{loc}}. \qquad (1.13)$$

We can evade the no-go result if this term is chosen to be negative. In what follows, we will discuss which objects contribute negatively and therefore can be used to satisfy equation (1.13).

In addition to the perturbative closed string sector, type II theories admit non-perturbative objects, the so-called Dirichlet p-branes (Dp-branes or D-branes in brief) and the orientifold planes (or simply O-planes). D-branes are objects, on which open strings can end. Strings can have both ends on the same D-brane; they can stretch between two different D-branes, or propagate (as closed strings) from one D-brane to another one. Due to their intrinsic tension, stretched strings give rise to massive excitation modes. When the distance between two or more branes is vanishing, the strings allow for massless modes. Hence, massless fields are localized at the intersection locus of branes. D-branes are classified according to their spatial dimension denoted by p.

For example, a D7-brane can be setup such that it fills the four-dimensional spacetime entirely – this accounts for the three spatial dimensions x_1, x_2, and x_3 (see table 1.2) – and wraps a four-dimensional subspace of the CY manifold (a holomorphic four-cycle extended, for instance, along the directions x_4, x_5, x_6, and x_7). The directions x_8 and x_9 are transverse to the brane. Stable configurations of D-branes underlie certain supersymmetric calibration conditions (BPS conditions). In type IIB, D-branes have to wrap complex subspaces of the CY manifold. Configurations of other kinds turn out to be unstable. D-branes, like strings, have

	x_0	x_1	x_2	x_3	x_4	x_5	x_6	x_7	x_8	x_9
D7	×	×	×	×	×	×	×	×		
D5	×	×	×	×	×	×				
D3	×	×	×	×						
E3						×	×	×	×	

Table 1.2: Spacetime extension of D-branes. Spacetime directions filled by the branes are denoted by crosses, whereas the transverse ones are left blank.

tension; stability is preserved only when they minimize this tension. Holomorphic cycles have minimal volume in their homology class. Therefore, type IIB compactifications naturally come with D3-, D5-, and D7-branes that wrap, respectively, holomorphic zero-cycles, two-cycles, and four-cycles of the internal space.

Consider the Bianchi identity (BI) for $\tilde{F}_{(5)}$:

$$d\tilde{F}_{(5)} = H_{(3)} \wedge F_{(3)} + 2\kappa^2 T_3 \rho_3 , \tag{1.14}$$

where T_3 is the D3-brane tension and ρ_3 is the local D3-charge density on the compact space. The latter enters the equation because $F_{(5)} = dC_{(4)}$ couples to D3-branes and O3-planes. The integration of the BI over the compact space gives the tadpole cancellation condition

$$\frac{1}{2\kappa^2 T_3} \int H_{(3)} \wedge F_{(3)} + N_3 = 0 . \tag{1.15}$$

This equation tells that the amount of contribution from the fluxes has to be compensated by the total D3-charge from local sources. The Dirac flux quantization requires $H_{(3)}$ and $F_{(3)}$ to be integer forms. Hence, formula (1.15) is a condition on three integers and thus admits only a discrete family of solutions.

We combine Einstein equations and the integrated BI into one condition. First, rewrite equation (1.15) in terms of $\alpha(y)$ and $G_{(3)}$ by making use of (1.11); then, subtract this identity from (1.13). We obtain, in this way, the main result of GKP's work:

$$\tilde{\nabla}^2 \left(e^{4A} - \alpha\right) = e^{2A} \frac{|iG_{(3)} - *G_{(3)}|^2}{24 \operatorname{Im} \tau} + e^{-6A} |\partial \left(e^{4A} - \alpha\right)|^2 + 2\kappa^2 e^{2A} \left(\frac{1}{4} T_{loc} - T_3 \rho_3\right) , \tag{1.16}$$

where the Hodge star operator is with respect to the metric of the internal manifold. What can we learn from this equation? First, assume

$$\frac{1}{4} T_{loc} - T_3 \rho_3 \geq 0 . \tag{1.17}$$

This condition restricts the choice of local sources. D3-branes and O3-planes, as well as D7 and O7 saturate the inequality; anti-D3-branes satisfy it; whereas all other objects (O5, anti-O3 etc.) violate this condition. Second, for equation (1.16) to hold, the terms on the rhs have to vanish separately each of them being positive. Hence, only those local sources that saturate (1.17) can be solutions of the eom. Note that this result holds at tree-level; indeed taking into proper account quantum effects would allow the inclusion of anti-D3-branes. Further, the three-form needs to be imaginary self-dual (ISD):

$$*G_{(3)} = iG_{(3)}. \tag{1.18}$$

This implies a relation between the warp factor and the self-dual five-form $\tilde{F}_{(5)}$ via

$$e^{4A} = \alpha(y). \tag{1.19}$$

Later on, in chapter 6, we will study the important role that the ISD property plays in stabilizing complex structure moduli and axio-dilaton.

Equations (1.15), (1.13) and (1.16) do not determine all eom's. To complete the set of conditions, the field equations for the internal components of the Ricci tensor (without the warp factor) and axio-dilaton remain to be determined. We state them without derivation [11]:

$$\tilde{R}_{mn} = \kappa^2 \frac{\partial_m \tau \partial_n \bar{\tau} + \partial_n \tau \partial_m \bar{\tau}}{4\,(\mathrm{Im}\tau)^2} + \kappa^2 \left(T^{D7}_{mn} - \frac{1}{8} \tilde{g}_{mn} T^{D7} \right),$$
$$\tilde{\nabla}^2 \tau = \frac{\tilde{\nabla}\tau \cdot \tilde{\nabla}\tau}{i\,\mathrm{Im}\tau} - 4\kappa^2 \frac{(\mathrm{Im}\tau)^2}{\sqrt{-g}} \frac{\delta S^{D7}}{\delta \bar{\tau}}. \tag{1.20}$$

Here S^{D7} is the action and T^{D7} the energy momentum tensor of the D7-brane. The Ricci tensor is not vanishing in general, hence the internal geometry does not need to be CY.

Models without D7-branes dramatically simplify the field equations for the internal geometry:

$$\tilde{R}_{mn} = 0, \qquad \partial_m \tau = 0. \tag{1.21}$$

We end up with a constant axio-dilaton. In this case, the compactification manifold X turns out to be conformally CY because of the warp factor e^{2A}. Finally, note that D7-branes would require the presence of O7-planes to saturate the tadpole. In this case, the resulting geometry would be a CY orientifold and hence no more CY.

GKP have outlined a clear strategy to find warped solutions. First, choose an internal manifold satisfying (1.20). Second, consider localized objects that saturate the bound (1.16). Third, the flux configuration needs a five-form flux $\tilde{F}_{(5)}$ defined in (1.6) with an ISD complex internal three-form flux $G_{(3)}$ from (1.18). The resulting solutions do not need to be supersymmetric ones. Indeed, it can be shown that equation (1.18) only implies that $G_{(3)} \in H^{(2,1)} \oplus H^{(0,3)}$, whereas supersymmetric solutions require in addition the $(0,3)$ contribution to vanish. We will discuss this point in more detail in chapter 6; see for instance expression (6.18).

1.2.3 Intersecting D7-branes

Until now we considered only fluxes from background closed strings. The presence of D7-branes charged under various gauge groups is an important ingredient for string model building in the type IIB context. They account for gauge groups, chiral matter and Yukawa couplings. But all this comes with a cost: their presence makes the construction of solutions more challenging. They contribute to the D7-charge tadpole that needs to be saturated by the introduction of

O7-planes. Furthermore, turning on fluxes on the D-branes induces D3-charges that contribute to (1.15). We address this issue in chapter 4, in which explicit type IIB flux compactification models are analyzed with a focus on the chirality problem arising from intersecting charged D-branes. In particular, see appendices A.1.1 and A.1.2 for details on the computation of D-brane and O-plane charges. The rest of this section presents a rough sketch of how the matter content arises from intersecting stacks of branes.

The excitation modes of a D-brane are open strings. A careful analysis of their spectrum shows that the massless modes give rise to the $U(1)$ gauge theory on the D-brane. Consider a bunch of N D7-branes on top of each other, this configuration is often referred to as a stack (of N D-branes). Then the massless modes induce a non-abelian group, more precisely a dimensionally reduced supersymmetric Yang-Mills (SYM) gauge theory. Because of supersymmetry, the gauge bosons come with fermionic superpartners. Both fields are in the adjoint representation of $U(N)$. Two four-cycles generically intersect each other in a one-dimensional complex subspace, i.e. a Riemann surface. Let D_1 be a four-cycle, on which a stack of N_1 D7-branes is wrapped giving rise to a $U(N_1)$ SYM gauge theory. Consider a second stack on a four-cycle D_2 (and the corresponding $U(N_2)$ gauge group) such that D_1 and D_2 intersect each other. It can be shown that chiral matter is induced at the intersection by bifundamental open strings 'stretched' between D_1 and D_2 with representation $(N_1, \bar{N}_2)$ or $(N_2, \bar{N}_1)$, depending on their orientation. Furthermore, three four-cycles generally intersect on three curves in the compact space. Where these three curves meet, chiral fermions from two different curves interact giving rise to Yukawa couplings among them. In conclusion, the type IIB compactification (with D-branes and orientifold planes) provides an efficient way to encode the entire matter content of MSSM-like theories in terms of the geometry of complex subspaces. Beside the massless modes arising from the perturbative sector in ten dimensions, we have: a Yang-Mills gauge theory on an eight-dimensional manifold (e.g., on a stack D_1), chiral matter on a six-dimensional manifold (e.g., $D_1 \cap D_2$), and Yukawa couplings in the four-dimensional spacetime (e.g., $D_1 \cap D_2 \cap D_3$).

1.3 F-theory

In this section, we give an overview of those aspects of F-theory that are relevant for our purposes. We do not aim at giving a complete coverage of this vast topic. For a comprehensive introduction to F-theory model building see for example [12, 13].

We should start arguing about the necessity of a strongly coupled description of type IIB strings. Once 7-branes (i.e. complex codimension one charged objects) enter the picture, a serious attempt to treat their backreaction is unavoidable. F-theory is a non-perturbative description of type IIB theory with 7-branes and varying axio-dilaton.

What we want to discuss in brief is the relation between F-theory and type IIB orientifolds with O7/O3-planes. We will see that the GKP solutions we derived in the previous sections are the supergravity and weak coupling limit approximation of solutions of F-theory models. More generally, Sen showed that any F-theory compactification on a CY fourfold admits a type IIB orientifold approximation in an appropriate weak coupling limit [14].

The standard approach of perturbative type IIB in presence of orientifolds and D-branes is to treat them as probe objects. It seems reasonable that asymptotically away from these objects we can neglect their backreaction on the background geometry, but care is needed. The backreaction of a stack of D-branes was first worked out in [15]. The main idea goes as follows. Dp-branes are sources of R-R background fields C_{p+1} in the $9-p$ normal directions. Poisson-type equations describe electrically charged objects:

$$\nabla^2 \phi(r) = \delta(r) \quad \Longrightarrow \quad \phi \sim \frac{1}{r^{n-2}}. \tag{1.22}$$

The harmonic function ϕ governs the profile of the electric potential sourced by a codimension n object, a D$(9-n)$-brane for instance. We see that the solution makes sense only for codimensions greater than two. A D7-brane has codimension two and hence is a critical object. Its spacetime position is completely determined by the two transverse coordinates x_8 and x_9 (see table 1.2) that can be combined into the complex variable $z = x_8 + ix_9$. Consider a D7-brane placed at z_0 in the complex plane normal to the brane. Its Poisson equation is

$$d * F_9 = \delta^{(2)}(z - z_0). \tag{1.23}$$

Integration of both sides of the equation yields

$$\int_C d * F_9 = \int_C dF_1 = \oint_{S^1} F_1 = \oint_{S^1} dC_0 = 1, \tag{1.24}$$

where we used the electro-magnetic duality (1.4). The solution of the Poisson equation scales logarithmically. For further purposes it is useful to state the solution in terms of the axio-dilaton; it can be show that, in the vicinity of a D7-brane, it is

$$\tau(z) = \tau(z_0) + \frac{1}{2\pi i} \ln(z - z_0) + \text{regular terms}. \tag{1.25}$$

There are two main characteristics of this solution. First, the string coupling approaches zero at the position of the brane, whereas it inevitably increases away from it. Let us recast equation (1.25) in a form that explicitly relates to the string coupling [16, 17]:

$$\operatorname{Im} \tau \simeq -\frac{1}{2\pi} \ln \left| \frac{z - z_0}{\lambda} \right|. \tag{1.26}$$

Here, λ is the overall scaling of the axio-dilaton, and it is related to τ_0. At the point $z - z_0 = \lambda$ the string coupling diverges: $g_s = 1/\operatorname{Im} \tau \longrightarrow \infty$. Hence, the phase of λ determines a special direction in the complex plane. The solution is not rotational invariant as might naively be expected. Nevertheless, if λ is chosen to be very large the rotational symmetry is approximately restored. In particular, the limit $\lambda \longrightarrow \infty$ extends the region in which the string coupling can be assumed to be weak. This is the so-called Sen's limit, i.e. a perturbative limit in which F-theory can be approximated by effective supergravity. We will return to this later.

Furthermore, formula (1.25) exhibits a logarithmic branch cut. Going around z_0 in the complex plane, the axio-dilaton transforms in a discontinuous way:

$$\tau \longrightarrow \tau + 1. \tag{1.27}$$

This monodromy is a global feature. Even if at asymptotic distances from the D7-brane spacetime looks locally flat, the geometry exhibits a deficit angle [16, 17]. The backreaction of the D7-brane cannot be neglected anymore, contrary to the case of objects of codimension greater than two. If we want to end up with a consistent type IIB theory containing D7-branes we have to deal with the fact that the string coupling tends to vary in a non-trivial way. Or equivalently, we have to develop a framework that incorporates the variation of the axio-dilaton field.

We start with the following observation. Action (1.5) is manifestly invariant under fractional linear transformations of the axio-dilaton and under matrix transformations of the R-R and NS-NS two-forms (recast in the form of a two-vector):

$$\tau \longrightarrow \frac{a\tau + b}{c\tau + d}, \qquad \begin{pmatrix} C_{(2)} \\ B_{(2)} \end{pmatrix} \longrightarrow \begin{pmatrix} a & b \\ c & d \end{pmatrix} \begin{pmatrix} C_{(2)} \\ B_{(2)} \end{pmatrix} = M \begin{pmatrix} C_{(2)} \\ B_{(2)} \end{pmatrix}, \qquad \det M = 1, \qquad (1.28)$$

with a, b, c, d real numbers. These transformations describe the $SL(2, \mathbb{R})$ group. It is a continuous symmetry of the theory in the supergravity approximation. This invariance is broken to $SL(2, \mathbb{Z})$ once non-perturbative effects are taken into account. These arise, for instance, when D-branes enter the model. Full type IIB superstring theory has $SL(2, \mathbb{Z})$ symmetry. Indeed, note that expression (1.27) is a special case of (1.28): $a = b = d = 1$ and $c = 0$. Hence, the monodromy induced by the D7-brane is consistently inherited in the symmetry of the theory.

One can make the $SL(2, \mathbb{Z})$ symmetry explicit by incorporating it in a geometrical way: the axio-dilaton can be interpreted as the complex structure of a two-torus. The torus accounts for two extra dimensions; it is fibered over each point of the ten-dimensional spacetime. Hence, the compactification manifold becomes eight-dimensional. Unfortunately, the resulting twelve-dimensional framework is not a good candidate for a theory of fundamental interactions. Even though it is referred to as F-theory, it should rather be understood as an auxiliary theory describing type IIB solutions with varying axio-dilatons. Indeed, there is no twelve-dimensional supergravity that can play the role of the low-energy limit of F-theory.

Even if there would be such a description, we would have a further puzzle. We postulate the existence of a four-form flux $G_{(4)}$ in the eight-dimensional internal space, whose reduction along the two one-cycles of the torus accounts for the presence of $H_{(3)}$ and $F_{(3)}$. But we run into difficulties in explaining why four- and two-forms do not appear in the type IIB action from the reduction of $G_{(4)}$ along a point and both one-cycles of the torus. Finally, it seems unnatural that only the complex structure of the two-torus does appear as a field in the supergravity action, but not its volume.

F-theory is dual to M-theory in the limit of a shrinking two-torus. A detailed description of the duality transcends the goal of this section. It suffices to know that there is a rigorous way (see for example [18]) to derive the effective theory in four dimensions on the F-theory side starting from M-theory compactification on $\mathbb{R}^{1,2} \times Y_4$, where Y_4 is an elliptic fibration over the complex threefold B_3:

$$\begin{array}{c} T^2 \xrightarrow{\iota} Y_4 \\ \downarrow \pi \\ B_3 \end{array} \qquad (1.29)$$

Each fiber is an elliptic curve, i.e. the zero locus of a complex cubic polynomial such that its points lie in a region topologically equivalent to a torus. In the limit where the elliptic fiber shrinks to zero volume, this is dual to F-theory on Y_4. This limit accounts for the fact that the volume of the torus does not appear in the field content of the effective theory. Furthermore, if the Y_4 is chosen to be Calabi-Yau, the effective theory describes an $\mathcal{N} = 1$ supersymmetric Minkowski vacuum in four dimensions.

F-theory compactified on Y_4 is type IIB on $\mathbb{R}^{1,3} \times B_3$ with the axio-dilaton being a function of the complex variables y_i parametrizing the base space B_3. More precisely, we choose $\tau(\vec{y})$ to transform as the modular parameter of the T^2 fibered over the base. A general complex elliptic curve can always be written in the Weierstrass form:

$$y^2 = x^3 + fx + g, \qquad (1.30)$$

where x and y are the complex coordinates of the curve, and $f, g \in \mathbb{C}$ are constants. The latter can be interpreted as deformation parameters determining the shape of the torus. We can construct an elliptic fibration by making them suitable polynomials in the base coordinates. More precisely, they should become sections of some powers of an appropriate line bundle:

$$f(\vec{y}) \in H^0\left(B_3, L^{\otimes 4}\right), \qquad g(\vec{y}) \in H^0\left(B_3, L^{\otimes 6}\right). \qquad (1.31)$$

Note that equation (1.30) needs to be homogeneous in order to be CY. This imposes further restrictions on the elliptic coordinates. In fact, they are sections themselves:

$$x \in H^0\left(B_3, L^{\otimes 2}\right), \qquad y \in H^0\left(B_3, L^{\otimes 3}\right). \qquad (1.32)$$

The parameter τ of the torus can be implicitly written in terms of the modular invariant equation:

$$j(\tau) = \frac{4(24f)^3}{\Delta}. \qquad (1.33)$$

Here, the denominator

$$\Delta = 4f^3 + 27g^2, \qquad (1.34)$$

is the discriminant of the elliptic equation. The fiber degenerates at the zero locus of Δ. This is a complex codimension one region. Furthermore, the modular function diverges as the torus modulus tends to the complex infinity:

$$\lim_{\tau \to i\infty} j(\tau) = \infty. \qquad (1.35)$$

This divergence is consistent with what we expect from the description of the axio-dilaton in the vicinity of a D7-brane (1.25). Note that the limit (1.35) still holds after the monodromy action (1.27). But, more generally, this result is invariant under $SL(2, \mathbb{Z})$ transformations of τ. These facts suggest to interpret the solutions of the discriminant equation as the locations of 7-branes. These are generalizations of D7-branes that appear only in F-theory and do not have perturbative counterparts in type IIB.

In conclusion, branes are located in the base space where the torus fibration becomes singular. The picture one should have in mind is the following. Imagine a life-belt-shaped torus attached at each point of the base manifold. Moving though B_3 the shape of the torus changes – according to (1.30). Along certain loci, one of the torus cycles shrinks to zero; we are left with a pinched life-belt. These loci are complex hypersurfaces and are determined by the zeros of the discriminant (1.34).

1.3.1 Sen's limit

Consider the following parametrization of the sections f and g:

$$f = -3h^2 + \varepsilon\eta, \qquad g = -2h^3 + \varepsilon h\eta + \varepsilon^2\chi, \qquad (1.36)$$

with h, η and χ being sections:

$$h \in H^0\left(B_3, L^{\otimes 2}\right), \qquad \eta \in H^0\left(B_3, L^{\otimes 4}\right), \qquad \chi \in H^0\left(B_3, L^{\otimes 6}\right), \qquad (1.37)$$

whereas ε is a constant that we wish to vary. There is no loss of generality in this representation of f and g, but it is redundant: we can arrange h, η, χ and ε to express f and g in different ways.

We can rewrite the discriminant in terms of the new sections, and expand Δ around $\varepsilon = 0$ up to second order terms:

$$\Delta = \varepsilon^2\left(-9h^2\eta^2 - 108h^3\chi\right) + \varepsilon^3\left(4\eta^3 + 54h\eta\chi\right) + 27\chi\varepsilon^4 \xrightarrow{\varepsilon \to 0} \varepsilon^2\left(-9h^2\right)\left(\eta^2 + 12h\chi\right) \quad (1.38)$$

The representation (1.36) has been chosen to cancel the zeroth and first order terms here. In this limit the discriminant locus splits into two complex codimension one hypersurfaces:

$$h = 0, \qquad \text{and} \qquad \eta^2 + 12h\chi = 0. \qquad (1.39)$$

By inserting this result in (1.33), we obtain

$$j(\tau) \simeq \frac{4\,(24)^3\left(\varepsilon\eta - 3h^2\right)^3}{\varepsilon^2\left(-9h^2\right)\left(\eta^2 + 12h\chi\right)}. \qquad (1.40)$$

The j-function diverges as ε goes to zero almost everywhere on the base manifold except for those regions of B_3 where the numerator vanishes; this happens where $|h| \sim \sqrt{|\epsilon|}$. Therefore, for small ε, j becomes large everywhere $|h| \gg \sqrt{|\epsilon|}$. In terms of the axio-dilaton field, this limit implies that $\tau \to i\infty$. This is the weak coupling limit of type IIB! A study [19] of the monodromies along contours encircling the zero loci of the discriminant (1.39) shows that, in the weak coupling limit, the axio-dilaton behaves in the same way as in the type IIB orientifold case where the O7-plane is located at $h = 0$ and the D7-branes at $\eta^2 + 12h\chi = 0$. Indeed, it can be shown that Sen's limit is the weak coupling limit of F-theory on Y that describes type IIB orientifold compactification on B_3.

We would like to take a closer look at the base manifold, and discuss whether we can say something about its structure. For this purpose, it is convenient to describe $\tilde{B}_3$ as the quotient space of an appropriate double cover X. This can be constructed by adding a new coordinate ξ, chosen to be an element of the line bundle L:

$$X: \quad \xi^2 = h(\vec{y}), \tag{1.41}$$

where $\vec{y}$ parametrizes the base manifold. Note that, if we exclude the zero locus of $h = 0$, for each point $\vec{y} \in B_3$ there are two points $(\vec{y}, \xi = \pm\sqrt{h})$. In this picture, the base space is given by the quotient $B_3 = X/\sigma$ with the orientifold involution

$$\sigma: \quad \xi \longrightarrow -\xi. \tag{1.42}$$

$h = 0$ is the $\mathbb{Z}_2$-invariant locus of this transformation.

We want to determine some properties of X, in particular, whether or not it is a CY threefold. Recall that h is a section of $L^{\otimes 2}$ and ξ a coordinate of L. Assuming that X is non-singular, then (1.41) is CY if and only if

$$c_1(X) = c_1(T_{B_3}) + c_1(L)(1-2) = 0. \tag{1.43}$$

Here, T_{B_3} denotes the tangent bundle of B_3. Since $c_1(K_{B_3}) = -c_1(T_{B_3})$, L needs to be the anti-canonical bundle of the base manifold in order to satisfy the condition above: $L = K_{B_3}^{-1}$. In the beginning, we assumed Y to be CY. By (1.29) and (1.30), we have then

$$c_1(Y) = c_1(T_{B_3}) + c_1(L)(3+2-6) = 0. \tag{1.44}$$

Here, we use the fact that x and y are sections of $L^{\otimes 2}$ and $L^{\otimes 3}$; the negative coefficient accounts for the Weierstrass equation. This condition is identical with (1.43). It follows that if Y is CY (fourfold) then X is CY (threefold).

In conclusion, Sen's limit is the weak coupling limit of F-theory compactification on Y. It yields a type IIB orientifold construction on B_3:

$$\text{type IIB on} \quad \mathbb{R}^{1,3} \times \underbrace{X/\sigma \cdot (-1)^{F_L} \cdot \Omega}_{B_3}. \tag{1.45}$$

The base manifold is the quotient space of a CY threefold X with involution action σ given in (1.42). The O7-plane is placed in X along the invariant locus of the $\mathbb{Z}_2$ symmetry generated by σ. Furthermore, the orientation reversal Ω exchanges worldsheet left- and right-movers; the signum $(-1)^{F_L}$ acts on the Ramond sectors of the left-moving modes changing their sign.

Chapter 2

Aspects of toric geometry

In this chapter we give a brief overview of the construction of toric varieties and their subvarieties in terms of lattice polytopes. We furthermore fix some of the notation that we will need in the following chapters. The reader has a vast choice of existing literature on the subject, for example [20–22] and the very comprehensive [23], to name a few. In particular, [24] addresses the construction of Calabi-Yau hypersurfaces and complete intersections with a focus on issues related to string duality. For a more pedagogical approach see for instance [25].

2.1 The fan

Toric varieties can be thought of as generalizations of weighted projective spaces. We can construct a $(r-n)$-dimensional toric variety X in terms of r homogeneous coordinates, an exceptional set Z, and the group identification $(\mathbb{C}^*)^{r-n} \times G$:

$$X = \frac{\mathbb{C}^r - Z}{(\mathbb{C}^*)^{r-n} \times G}. \tag{2.1}$$

These building blocks are encoded in a fan Σ that completely determines X. The fan is a finite collection of strongly convex (i.e. they always have an apex) integral (i.e. they are spanned by lattice vectors) polyhedral cones with their apex in the origin such that the following conditions are satisfied: 1) any face of a cone $\sigma \in \Sigma$ belongs to Σ; and 2) given two cones $\sigma, \tau \in \Sigma$, their intersection is again contained in Σ. Note that in general σ and τ may have different dimensions. The *n-skeleton* $\Sigma(n) \subset \Sigma$ denotes the set of n-dimensional cones. Consider the rays $\rho_j \in \Sigma(1)$. Each of them is generated by an integral vector v_j (also called *primitive vector*) in a n-dimensional lattice, which we denote by N. v_j spans from the origin towards the nearest point of the lattice along the direction of ρ_j. To each primitive vector v_j we associate a homogeneous coordinate z_j and a divisor

$$D_j = \{z \in X : z_j = 0\}. \tag{2.2}$$

The group $(\mathbb{C}^*)^{r-n}$ is determined by the $r-n$ weighted scalings ($i = 1, \ldots, r-n$):

$$(z_1 : \ldots : z_r) \longrightarrow (\lambda^{w_{i1}} z_1 : \ldots : \lambda^{w_{ir}} z_r) \quad \text{with} \quad \sum_{j \leq r} w_{ij} v_j = 0 \in N \quad \text{and} \quad \lambda \in \mathbb{C}^*, \tag{2.3}$$

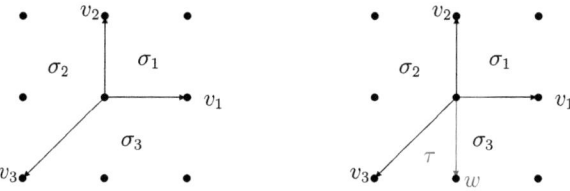

Figure 2.1: The fans of $\mathbb{P}^2$ and $\mathbb{F}_1$.

where w_{ij} are the entries of a $r \times (r - n)$ matrix we refer to as *weight matrix*. Moreover, $G \cong N/\text{span}(v_1, \ldots, v_r)$ is a finite abelian group, which accounts for phase symmetries. It arises if the one-skeleton does not span the entire lattice N. For example, let us consider a lattice $\hat{N}$ that is completely spanned by $\Sigma(1)$. Consider a refinement $N \supset \hat{N}$ such that $N \neq \text{span}(v_1, \ldots, v_r)$. Then we have $G \cong N/\hat{N}$. Furthermore, the fan determines the exceptional set Z. This is the set of invariant points under the continuous group identification. A subset of coordinates is allowed to vanish simultaneously, i.e. $z_{j_1} = \ldots = z_{j_k} = 0$ (or equivalently $D_{j_1} \cdot \ldots \cdot D_{j_k} \neq 0$), iff there is a cone that contains the corresponding rays $\rho_{j_1}, \ldots, \rho_{j_k} \subset \sigma$. The exceptional set is the union of sets Z_I with minimal index sets I of rays for which there is no cone that contains them: $Z = \bigcup_I Z_I$.

Example 1. $\mathbb{P}^2$ **and its blow-up to** $\mathbb{F}_1$

We start with the second most simple example of a toric variety (the first being $\mathbb{P}^1$).[1] Consider the two-dimensional projective space. It is defined as

$$\mathbb{P}^2 = \frac{\mathbb{C}^3 - \{z_1 = z_2 = z_3 = 0\}}{(z_1 : z_2 : z_3) \sim (\lambda z_1 : \lambda z_2 : \lambda z_3)}, \quad \lambda \in \mathbb{C}^*. \quad (2.4)$$

The three-dimensional complex space accounts for the three homogeneous coordinates z_i. The complex space is modded out by a linear relation among these coordinates, which is displayed in the denominator. Three coordinates and an equivalence yield a two-dimensional variety. Note that the only fix point of the $\mathbb{C}^*$-identification is $(0:0:0)$; but the origin is here the exceptional set Z of definition (2.1). Its subtraction eliminates the fix point and makes the variety smooth.[2] Even worse, if we had not subtracted the origin in this case, the resulting variety would have not even been Hausdorff!

The lhs of figure 2.1 shows the fan of $\mathbb{P}^2$. It contains seven cones. The one-dimensional cones are spanned by the generators

$$v_1 = \begin{pmatrix} 1 \\ 0 \end{pmatrix}, \quad v_2 = \begin{pmatrix} 0 \\ 1 \end{pmatrix} \quad \text{and} \quad v_3 = \begin{pmatrix} -1 \\ -1 \end{pmatrix}, \quad (2.5)$$

---

[1] Well, it is matter of taste what we call "the simplest" here. Indeed, an even simpler toric variety would be the punctured complex plane $\mathbb{C}^*$; let us refer to this as "the zeroth" most simple example.

[2] In general, the subtraction of the exceptional set is not sufficient to ensure that the resulting variety is non-singular. In particular, singularities can still arise from non-basic subdivisions of the fan; see subsection 2.1.1.

represented in an appropriate basis of N. The pairs $\{v_1, v_2\}$, $\{v_2, v_3\}$ and $\{v_3, v_1\}$ span the two-dimensional cones σ_1, σ_2 and σ_3, respectively. The origin is the only zero-dimensional cone. The relation among the generators v_i is

$$1\, v_1 + 1\, v_2 + 1\, v_3 = 0\,. \tag{2.6}$$

To this relation we associate the parameter $\lambda \in \mathbb{C}^*$. The coefficients of the generators give the exponents of λ in expression (2.1). For each generator v_i there is a homogeneous coordinate z_i. This, in turn, determines a divisor $D_i = \{z_i = 0\}$. Furthermore, note that v_1, v_2 and v_3 do not share any cone. This fact identifies the exceptional set correctly as $Z = \{z_1 = z_2 = z_3 = 0\}$.

Consider a slight modification of the $\mathbb{P}^2$ fan. Let us add an additional generator with lattice coordinates $w = (0, -1)^T$, denoted with red color on the rhs of figure 2.1. We can recover the data of the variety completely by starting from its associates fan. The additional generator splits the cone σ_3 of $\mathbb{P}^2$ into two cones. The resulting fan contains nine cones. Moreover, the exceptional set is enhanced and has two elements. In addition to (2.6), we have now a second relation among generators:

$$v_2 + w = 0\,. \tag{2.7}$$

This relation accounts for a further parameter $\mu \in \mathbb{C}^*$ that enters in the definition of the variety:[3]

$$\mathbb{F}_1 = \frac{\mathbb{C}^4 - \{z_1 = z_3 = 0,\, z_2 = w = 0\}}{(z_1 : z_2 : z_3 : w) \sim (\lambda z_1 : \lambda\mu z_2 : \lambda z_3 : \mu w)}\,. \tag{2.8}$$

This is again a two-dimensional manifold (four coordinates and two relations). It is the Hirzebruch surface $\mathbb{F}_1$. Consider now the following two cases: first, let $w \neq 0$. In this case, we can rescale the homogeneous coordinates with $\mu = 1/w$. This yields all points $(z_1 : z_2 : z_3 : 1) \equiv (z_1 : z_2 : z_3)$ of the original $\mathbb{P}^2$ except for $(0 : 1 : 0)$. The latter has been excluded by the subdivision of the cone spanned by $\{v_1, v_3\}$. Consider the second case, $w = 0$; this implies $z_2 \neq 0$ since $\{z_2 = w = 0\}$ is an element of the exceptional set. Therefore, we can scale $z_2 = 1$ and hence obtain $(z_1 : 1 : z_3 : 0)$. Note that the point $(0 : 1 : 0)$ has been replaced by a $\mathbb{P}^1$ with the homogeneous coordinates $(z_1 : z_3)$. Deforming a given space by replacing one of its points by a $\mathbb{P}^1$ is called blow-up. Blow-ups are useful for desingularizing varieties; we will return on this later. Torically, a blow-up corresponds to the addition of a one-dimensional cone to the fan. Indeed, the fan on the rhs of figure (2.1) is the blow-up of the fan on the lhs. Blow-ups of $\mathbb{P}^2$ at up to eight generic points correspond to the so-called del Pezzo surfaces. These are two-dimensional Fano varieties, i.e. non-singular complex spaces with ample anti-canonical bundle; they play a prominent role in string model building as we shall see in chapters 4 and 5. Del Pezzos are classified according to the number of blow-ups of the two-dimensional projective space. In particular, the first Hirzebruch surface is a del Pezzo surface of type 1, denoted by dP_1. The $\mathbb{P}^2$ is a del Pezzo itself: the dP_0, corresponding to the zeroth blow-up.

---

[3]With an abuse of notation, w denotes the new generator as well as the corresponding homogeneous coordinate.

z_1	z_2	z_3	w
1	1	1	1
0	1	0	1

Table 2.1: Weight matrix of $\mathbb{F}_1$.

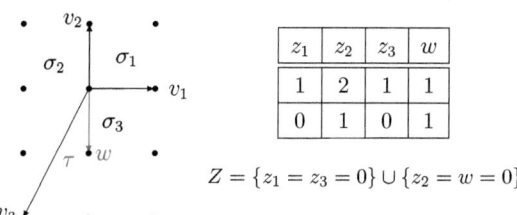

$$Z = \{z_1 = z_3 = 0\} \cup \{z_2 = w = 0\}$$

Figure 2.2: The fan of the Hirzebruch surface $\mathbb{F}_2$ and its weight matrix.

Table 2.1 shows the weight matrix of $\mathbb{F}_1$. Each line corresponds to an identification relation of the form (2.3). The Hirzebruch surface is described by four homogeneous coordinates and two such relations, hence the weight matrix is 2×4. Consider the first line of weights excluding the coordinate w. These are the exponents that determine the weighted scaling of the $\mathbb{P}^2$. The second line represents the $\mathbb{P}^1$.

2.1.1 On the singularity and compactness of a toric variety

X_Σ is *compact* iff its fan is complete, i.e. if Σ spans $N_\mathbb{R}$.[4] Both fans of figure 2.1 are complete, hence $\mathbb{P}^2$ and $\mathbb{F}_1$ are both compact varieties. But consider, for instance, a fan whose cones are contained in the quadrant of positive integers $\mathbb{N}^2 \subset N = \mathbb{Z}^2$. In this case, Σ has only four cones: the origin, ρ_1, ρ_2 and σ_1; this fan is not complete, hence the variety non-compact.

There are only two sources of singularity of toric varieties. Indeed, X_Σ is non-singular iff its fan satisfies both of the following conditions: a) it is *basic*, i.e. any cone $\sigma \in \Sigma$ is spanned by a subset of the basis of the lattice N; b) it is *simplicial*, i.e. all cones $\sigma \in \Sigma$ are simplices. Clearly, $\mathbb{P}^2$ and $\mathbb{F}_1$ are smooth varieties.

Example 2. The Hirzebruch surface $\mathbb{F}_2$

We present an example of a singularity induced by a cone that is not basic, and give its resolution. Consider the following weighted projective space:

$$\mathbb{P}_{121} = \frac{\mathbb{C}^3 - \{z_1 = z_2 = z_3 = 0\}}{(\lambda z_1 : \lambda^2 z_2 : \lambda z_3)}. \tag{2.9}$$

---

[4] $N_\mathbb{R} = N \otimes_\mathbb{Z} \mathbb{R} \cong \mathbb{R}^n$ is the real extension of the n-dimensional lattice N.

This looks similar to $\mathbb{P}^2$ beside of the fact that the scaling weight of the second homogeneous coordinate is 2 instead of 1 – cfr. (2.4). $\mathbb{P}_{121}$ is clearly singular, its fan (see figure 2.2) has a non-basic cone: the cone spanned by $\{v_1, v_3\}$ has volume 2.[5] Indeed, the point $(0 : 1 : 0)$ is fixed under the action

$$z_2 \longleftrightarrow -z_2, \tag{2.10}$$

This describes a $\mathbb{Z}_2$-quotient singularity.

The Hirzebruch surface $\mathbb{F}_2$ is the blow-up of $\mathbb{P}_{121}$:

$$\mathbb{F}_2 = \frac{\mathbb{C}^4 - \{z_1 = z_3 = 0, z_2 = w = 0\}}{(\lambda z_1 : \lambda^2 \mu z_2 : \lambda z_3 : \mu w)}. \tag{2.11}$$

Its fan is characterized by the addition a new generator w, which splits the cone $\{v_1, v_3\}$ in two parts: $\{v_1, w\}$ and $\{w, v_3\}$.

The conifold is an example of singularity caused by a non-simplicial cone. This is a well studied non-compact toric variety. We refer the interested reader to [24] for a detailed review of this singularity with a focus on its toric resolutions.

2.2 Line bundles and toric divisors

In this section, we explain the relation between line bundles and divisors. We focus on the description of toric divisors. As we will see, some of their properties can be studied in terms of the combinatorics of their associated polyhedrons.

2.2.1 The Picard and the divisor group

Let us start with a small step, by reminding the reader of the definition of a complex line bundle. A holomorphic line bundle is a holomorphic vector bundle whose fibers are complex one-dimensional vector spaces. Let X be a complex manifold with an open covering $X = \bigcup_\alpha U_\alpha$. A line bundle L on X is topologically described in each coordinate patch U_α by the product space $U_\alpha \times \mathbb{C}$. Its global structure is completely determined in terms of the transition functions $\{g_{\alpha\beta}\}$, which determine how the fibers combine in the non-zero overlap of two patches, $U_\alpha \cap U_\beta$. These functions need to be holomorphic, non-zero and finite, and furthermore to satisfy the cocycle conditions:

$$g_{\alpha\beta} g_{\beta\alpha} = g_{\alpha\beta} g_{\beta\gamma} g_{\gamma\alpha} = 1. \tag{2.12}$$

Holomorphic line bundles have a natural group structure with respect to the tensor product. The product of two bundles is the bundle characterized by the product of their transition functions. Moreover, the inverse element is called the dual line bundle:

$$L \otimes L' \sim \{g_{\alpha\beta} \cdot g'_{\alpha\beta}\}, \qquad L^* \sim \{g_{\alpha\beta}^{-1}\}, \qquad L \otimes L^* = \text{identity}. \tag{2.13}$$

---

[5]More precisely, the *basic simplex* constructed over this cone has lattice-volume 2. This simplex is constructed by bounding the cone with the hyperplane on which lie the generators of that cone.

These operations define an abelian group, the so-called *Picard group* of holomorphic line bundles on X, which is denoted by Pic(X). We now turn our attention to the geometrical interpretation of line bundles. In particular, we want to discuss the relation between the Picard group and a special class of subspaces of X, which are called Cartier divisors.

Given an open covering $X = \bigcup_\alpha U_\alpha$, a *Cartier divisor* is a codimension one subvariety $D \subset X$ defined by rational local functions f_α, such that $g_{\alpha\beta} = f_\alpha/f_\beta$ is regular and non-zero on the overlap of coordinate patches $U_\alpha \cap U_\beta$. Any such meromorphic function can always be written as a quotient:

$$f_\alpha = \frac{g_\alpha}{h_\alpha}, \tag{2.14}$$

with h_α and g_α holomorphic functions on the coordinate patch U_α.

Consider a hypersurface V described locally by f_α. The vanishing order of f_α in a neighborhood of $x \in V$ is the difference between the orders of g_α and h_α:

$$\gamma = \mathrm{ord}_{V,x}(f_\alpha) = \mathrm{ord}_{V,x}(g_\alpha) - \mathrm{ord}_{V,x}(h_\alpha). \tag{2.15}$$

If γ is positive, the associated function has zeros of order γ; vice versa, if it is negative, the function has poles of order $-\gamma$. If V is irreducible,[6] it can be shown [26] that (2.15) is independent on x, and hence on the patch. Therefore, any Cartier divisor can be written as a sum over irreducible hypersurfaces $D_j \subset X$:

$$(f) = \sum_j \mathrm{ord}_{D_j}(f) D_j = \sum_j \mathrm{ord}_{D_j}(g_j) D_j - \sum_j \mathrm{ord}_{D_j}(h_j) D_j. \tag{2.16}$$

This is the difference of two effective divisors: the *zero divisor* and the *pole divisor*. From this expression we can see that any Cartier divisor is also a Weil divisor. A *Weil divisor* is a codimension one subvariety of a complex variety X defined by the locally finite[7] formal sum

$$D = \sum_j a_j D_j, \tag{2.17}$$

where the D_j's are irreducible hypersurfaces – also called prime divisors – and a_j's integer coefficients. The divisor is called effective if $a_j \geq 0$. Weil divisors can be added, subtracted and there is a clear definition of the inverse element. They thus form an additive *divisor group* that we denote by Div(X). Compare expression (2.16) with (2.17). Clearly, any Cartier divisor defines a Weil divisor, whose coefficients a_j are the orders of the zeros (if positive) or poles (if negative) of f along the prime divisors D_j. It turns out that on a smooth ambient variety, any Weil divisor is also Cartier.

Cartier divisors are naturally associated with holomorphic line bundles. Indeed, the rational functions f_α/f_β satisfy the cocycle conditions (2.12) and can be though of transition functions $g_{\alpha\beta}$. Hence, they define a holomorphic line bundle: $L = \mathcal{O}(D)$. In this picture, the collection

---

[6]A divisor is irreducible if it can not be written as the union of two distinct hypersurfaces.

[7]Consider a neighborhood U_p around an arbitrary point $p \in X$. Then only finitely many irreducible hypersurfaces D_j intersect U_p. Furthermore, if X is compact then locally finiteness implies finiteness, i.e. the sum is finite.

of all local functions $\{f_\alpha\}$ determine a global meromorphic section of L. Cartier divisors form a linear group that we denote by $\mathrm{CDiv}(X)$. Addition and subtraction of divisors correspond to multiplication and division of their associated sections:

$$D + D' \sim f \cdot f', \qquad D - D' \sim f/f'. \tag{2.18}$$

Note that the quotient p of two meromorphic sections f and f' of the same bundle is a rational function:

$$p_\alpha = \frac{f_\alpha}{f'_\alpha} = \frac{g_{\alpha\beta} f_\beta}{g_{\alpha\beta} f'_\beta} = \frac{f_\beta}{f'_\beta} = p_\beta. \tag{2.19}$$

Rational functions do not change the line bundle. Indeed, the transition functions of the quotient section are all equal to one, hence the line bundle trivial. The associated divisor (p) is called a *principal divisor*. Two divisors D and D' are defined to be linearly equivalent if they differ by a principal divisor (p):

$$D \equiv D' \quad \Longleftrightarrow \quad D = D' + (p), \tag{2.20}$$

where p is a meromorphic function on X. We say that D and D' belong to the same *divisor class* modulo addition of principal divisors, which are those divisors associated to rational functions. These form a subgroup $\mathrm{CDiv}_p(X) \subset \mathrm{CDiv}(X)$.

This equivalence casts light on the relation between divisors and line bundles. Consider the sum of two divisors, the resulting line bundle is then given by the simple additive structure

$$\mathcal{O}(D + D') = \mathcal{O}(D) \otimes \mathcal{O}(D'). \tag{2.21}$$

Here, the map

$$\mathcal{O} : \mathrm{CDiv}(X) \longrightarrow \mathrm{Pic}(X) \tag{2.22}$$

is a surjective group homomorphism, because linearly equivalent divisors define the same line bundle:

$$D = D' + (p) \quad \Longleftrightarrow \quad \mathcal{O}(D) = \mathcal{O}(D'). \tag{2.23}$$

Principal divisors correspond to the identity element of the Picard group; hence, $\mathcal{O}((p))$ is trivial. It can be be shown that there is a group isomorphism between the quotient group of Cartier modulo principal divisors and the Picard group:

$$\mathrm{CDiv}(X) / \mathrm{CDiv}_p(X) \longrightarrow \mathrm{Pic}(X)$$
$$D \bmod (p) \longmapsto \mathcal{O}(D).$$

2.2.2 Polytopes and divisors

Consider a toric variety X defined by an expression of the type (2.1). Then, X is endowed with a particular simple class of hypersurfaces. To any homogeneous coordinate z_j we can associate the *toric divisor* characterized by the vanishing locus (2.2). Sums of toric divisors with respect to integral coefficients generate the Weil group:

$$\mathrm{Div}(X) = \bigoplus_{j=1}^{r} \mathbb{Z} D_j. \tag{2.24}$$

28

The group of Weil divisors modulo linear equivalences is called the *Chow group* of X:

$$\text{Div}(X)/\text{Div}_p(X) \cong A_{r-1}(X). \tag{2.25}$$

Here, $\text{Div}_p(X)$ is the subgroup of principal divisors of X. There are as many inequivalent divisors classes as weighted scalings, i.e. $r - n = \text{rank}(A_{r-1}(X))$.

Relevant properties of divisors can be rephrased in terms of the combinatorics between lattice points and cones [24, 27]. In order to show these relations, we first need to define the dual lattice to N as $M = \text{Hom}(N, \mathbb{Z})$ with the canonical pairing $\langle , \rangle$. Consider a toric variety X_Σ determined by the fan $\Sigma \subset N_\mathbb{R}$. Any point $m \in M$ of its dual (integral) lattice defines a principal divisor $(m) = \text{div}(\chi^m)$ as the zero locus of the rational equations

$$\chi^m = \prod_{j=1}^{r} z_j^{\langle m, v_j \rangle} = 0. \tag{2.26}$$

The Chow group is generated by the irreducible toric divisors D_j modulo divisors $\text{div}(\chi^m)$. It can be shown, that there are two short exact sequences relating the Cartier and the Weil divisors, as well as the Picard and the Chow group, by injective group homomorphisms:

$$\begin{array}{ccccccccc}
0 & \longrightarrow & M & \longrightarrow & \text{CDiv}(X) & \longrightarrow & \text{Pic}(X) & \longrightarrow & 0 \\
& & & & \downarrow \iota & & \downarrow \iota & & \\
0 & \longrightarrow & M & \longrightarrow & \text{Div}(X) & \longrightarrow & A_{r-1}(X) & \longrightarrow & 0
\end{array} \tag{2.27}$$

For singular toric varieties, the Picard group is a non-trivial subgroup of the Chow group. In general,

$$\text{rank}(\text{Pic}(X)) \leq \text{rank}(A_{r-1}(X)). \tag{2.28}$$

On the other hand, if X is smooth then $\text{CDiv}(X) \cong \text{Div}(X)$ and hence $\text{Pic}(X) \cong A_{r-1}(X)$. Furthermore, it can be shown that if X is simplicial (i.e. singularities can only arise from non-basic cones) then the Cartier divisors are integer multiples of the Weil divisors. Hence the Picard group is a finite index subgroup of the Chow group.

A Weil divisor D is Cartier if for any maximal-dimensional cone $\sigma \in \Sigma(n)$ there is a point $m_\sigma \in M$ such that the coefficient of the formal sum is $a_j = -\langle m_\sigma, v_j \rangle$ for all rays $\rho_j \in \sigma$. Moreover, to each Cartier divisor D we associate a *polytope* as follows[8]

$$\Delta_D = \{m \in M_\mathbb{R} : \langle m, v_j \rangle \geq -a_j \ \forall \ \rho_j \leq r\}, \tag{2.29}$$

where $M_\mathbb{R}$ is the real extension of M.

It can be shown that the integer multiple kD of a divisor class D is associated to an extension of the divisor's polytope:

$$\Delta_{kD} = k\Delta_D. \tag{2.30}$$

---

[8]More precisely, this formula defines a polytope even if the divisor is not Cartier. But for our purposes we can restrict the attention to Cartier divisors.

Further, the vertices of the polytope can be translated within M without affecting the divisor class:

$$\Delta_{D+\text{div}(\chi^m)} = \Delta_D - m \,. \tag{2.31}$$

This is a consequence of the fact that adding principal divisors $\text{div}(\chi^m)$ to D does not change neither the divisor class nor the associated line bundle.

The corresponding line bundle $\mathcal{O}(D)$ is determined by the sections

$$s_{\Delta_D} = \sum_{m \in \Delta_D \cap M} c_m \chi^m = \sum_{m \in \Delta_D \cap M} c_m \prod_j z_j^{\langle m, v_j \rangle} \,. \tag{2.32}$$

These are Laurent polynomials whose exponents are given by the canonical pairing between the vectors spanning the divisor polytope and the generators of the ambient fan. Note that the monomials are invariant under weighted scalings (2.3). Therefore, s_Δ defines a meromorphic function on the variety. If we restrict s_{Δ_D} to an affine patch U_σ with $\sigma \in \Sigma(n)$, we can define a regular function by the local section

$$f_\sigma = s_{\Delta_D}/\chi^{m_\sigma} \,. \tag{2.33}$$

But to define a hypersurface globally, we do not need functions. We need a polynomial equation as follows:

$$P_{\Delta_D} = \sum_{m \in \Delta_D \cap M} c_m \prod_j z_j^{\langle m, v_j \rangle + a_j} = 0 \,. \tag{2.34}$$

This polynomial is holomorphic and transforms homogeneously:

$$P_{\Delta_D}(\lambda^{w_i} z_i) = \lambda^{(\sum_{i=1}^r w_i)} P_{\Delta_D}(z_i) \,. \tag{2.35}$$

A Cartier divisor D is *base point free* iff $m_\sigma \in \Delta_D$ for all $\sigma \in \Sigma(n)$. Further, a Cartier divisor D is *ample* iff there is a bijection between vertices of Δ_D and $m_\sigma \in \Sigma(n)$.

2.3 Batyrev's construction of CY hypersurfaces

In what follows we would like to discuss the construction of CY manifolds as hypersurfaces embedded in toric ambient spaces. This construction is due to the seminal work of Batyrev [28]. Its power lies in the fact that topological quantities of the resulting CYs, such as their Hodge numbers, characteristic classes and intersection structures of toric divisor classes, are encoded in terms of the combinatorics of the associated polytopes. A sufficient and necessary condition for a polytope to describe a CY hypersurface is reflexivity. This property ensures the construction of mirror pairs of CY manifolds, and hence plays an important role in the study of mirror symmetry.

2.3.1 Reflexive polytopes

Given a fan Σ, the convex hull of the generators of the one-dimensional cones defines a unique polytope. Figure 2.3 shows the convex hull of the fan associated to the Hirzebruch $\mathbb{F}_1$. This

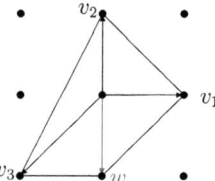

Figure 2.3: The convex hull of $\Sigma(\mathbb{F}_1)$.

hull describes a two-dimensional polytope. Conversely, given a polytope it is not clear which fan is associated to it. For example, the $\mathbb{P}_{121}$ and its blow-up, the Hirzebruch surface $\mathbb{F}_2$, have the same convex hull. Consider, instead, $\Delta \subset M$ defining an ample Cartier divisor. In this case, it can be shown that there is a uniquely associated fan to such a polytope: the normal fan Σ_Δ. This is the fan of cones over the faces of the dual polytope $\Delta^\circ \in N_\mathbb{R}$ defined by

$$\Delta^\circ = \{x \in N_\mathbb{R} : \langle m, x \rangle \geq -1 \ \forall \, m \in \Delta\} \,. \tag{2.36}$$

A lattice polytope whose dual is again a lattice polytope is called *reflexive*. Reflexivity is a very restrictive condition on polytopes. It implies $(\Delta \cap M_\mathbb{R}) \subset M$ and $(\Delta^\circ \cap N_\mathbb{R}) \subset N$. Indeed, prescription (2.36) relates to each vertex of Δ an equation describing the hyperplane bounding a facet of Δ°, but it is by no means trivial that all such hyperplanes intersect at integral loci of $N_\mathbb{R}$. Reflexive polytopes describe ample Cartier divisor classes. Let us denote by $\chi_\Delta \subset X$ a hypersurface associated to such a divisor class. Then, reflexivity is a sufficient and necessary condition for χ_Δ to have vanishing first Chern class, hence to be CY. Furthermore, it can be shown [28], that for an n-dimensional χ_Δ with $n \geq 3$, the Hodge numbers are given in terms of the combinatorics of the polytopes as follows:

$$h_{11}(\chi_\Delta) = h_{21}(\chi_{\Delta^\circ})$$
$$= l(\Delta^\circ) - 1 - \dim \Delta - \sum_{\text{codim}(\theta^\circ)=1} l^*(\theta^\circ) + \sum_{\text{codim}(\theta^\circ)=2} l^*(\theta^\circ) l^*(\theta) \,. \tag{2.37}$$

Here, θ and θ° denote a pair of dual faces of Δ and Δ°, respectively. $l(\theta^\circ)$ and $l^*(\theta^\circ)$ count the number of lattice points and interior lattice points of θ°. Mirror symmetry is manifest in this formula: χ_{Δ° is the mirror dual to χ_Δ. It can be constructed by exchanging the polytopes, i.e. by inserting Δ° into $N_\mathbb{R}$ and Δ into $M_\mathbb{R}$. Their Hodge numbers are related as follows:

$$h_{11}(\chi_\Delta) = h_{21}((\chi_\Delta)^*) = h_{21}(\chi_{\Delta^\circ})\,, \qquad h_{21}(\chi_\Delta) = h_{11}((\chi_\Delta)^*) = h_{11}(\chi_{\Delta^\circ})\,, \tag{2.38}$$

where here χ^* denotes the mirror dual to χ.

There are only 16 two-dimensional reflexive polytopes. Kreuzer and Skarke classified all three- and four-dimensional reflexive polytopes [29, 30]. There are 4 319 reflexive polytopes in three dimensions and 473 800 776 in four dimensions. There is no complete enumeration of reflexive polytopes in dimensions higher than four. Skarke estimates [31] their number to grow super-exponentially with the dimension d:

$$N_d \simeq 2^{2^{d+1}-4} \,. \tag{2.39}$$

31

According to this empirical formula there would be roughly 10^{18} and 10^{37} reflexive polytopes in five and six dimensions, respectively.

2.4 The intersection ring

Out of the almost half a billion reflexive polytopes in four dimensions, formula (2.37) yields 30 108 different Hodge data corresponding to 15 122 mirror pairs of CY threefolds. Simply connected CY threefolds are completely determined up to diffeomorphisms by their Hodge numbers, intersection rings and second Chern classes [32, 33]. In what follows, we would like to compute the intersection ring and the Chern classes of CY hypersurfaces embedded in toric ambient spaces.

The intersection ring of a compact and smooth toric variety X_Σ is the quotient ring [21]:

$$\frac{\mathbb{Z}[D_1,\ldots,D_r]}{\langle I_{SR}, I_{lin}\rangle}. \tag{2.40}$$

I_{SR} is the *Stanley-Reisner ideal*, it contains non-linear relations among divisors and is closely related to the exceptional set. Elements of this ideal have the following form:

$$D_{j_1}\cdot\ldots\cdot D_{j_k} = 0. \tag{2.41}$$

These are the intersections between those divisors whose associated homogeneous coordinates are elements Z_I of the exceptional set Z. Further, I_{lin} is the ideal generated by linear relations of the type

$$\sum_{j=1}^r \langle m, v_j\rangle D_j = 0. \tag{2.42}$$

These are closely related to the identification due to weighted scalings, or in terms of the fan, to those generators that sum to zero.

For any maximal-dimensional cone $\sigma \in \Sigma(n)$ that is simplicial, there is a simple formula for the intersection number of the divisors whose associated generators (let say, $v_{j_1},\ldots,v_{j_k}$) span σ:

$$D_{j_1}\cdot\ldots\cdot D_{j_k} = \frac{1}{\mathrm{Vol}(\sigma)}, \tag{2.43}$$

where $\mathrm{Vol}(\sigma)$ is the lattice-volume of the basic simplex of the cone. Note that if the toric variety is smooth then the volume of all basic simplices is equal to one.

Consider, for example, the Hirzebruch surface $\mathbb{F}_1$. Its Stanley-Reisner ideal contains two elements: $D_1 \cdot D_2 = 0$ and $D_3 \cdot D_w = 0$. Its linear ideal is given by the relations (2.6) and (2.7). The intersection ring is:

$$D_1\cdot D_2 = 1, \quad D_2\cdot D_3 = 1, \quad D_3\cdot D_w = 1, \quad D_w\cdot D_1 = 1. \tag{2.44}$$

The intersection ring of the weighted projective space $\mathbb{P}_{121}$ is:

$$D_1\cdot D_2 = 1, \quad D_2\cdot D_3 = 1, \quad D_3\cdot D_1 = \frac{1}{2}. \tag{2.45}$$

As one can see from figure 2.2, the cone spanned by $\{v_1, v_3\}$ has volume 2. According to formula (2.43), this accounts for the fractional intersection number between the divisors D_1 and D_3. Furthermore, the intersection number of the Hirzebruch surface $\mathbb{F}_2$ is just as it is expected from a non-singular variety:

$$D_1 \cdot D_2 = 1\,, \quad D_2 \cdot D_3 = 1\,, \quad D_3 \cdot D_w = 1\,, \quad D_w \cdot D_1 = 1\,. \tag{2.46}$$

Part of the intersection ring of an embedded hypersurface is inherited from the ring of the ambient space X. Let D describe the divisor class of the hypersurface, we obtain then

$$\begin{aligned} D_{j_1} \cdot \ldots \cdot D_{j_{k-1}}\Big|_D &= \int_D \text{PD}(D_{j_1}) \wedge \ldots \wedge \text{PD}(D_{j_{k-1}}) \\ &= \int_X \text{PD}(D_{j_1}) \wedge \ldots \wedge \text{PD}(D_{j_{k-1}}) \wedge \text{PD}(D)\,. \end{aligned} \tag{2.47}$$

Here, $PD(D)$ denotes the Poincaré form dual to D.[9] The ring computed in this way contains only intersections between torically induced divisors of the hypersurface. In chapter 3 we will present a computer assisted procedure aimed at constructing non-singular CY threefolds starting from reflexive polytopes, which computes their intersection rings and Chern classes.

---

[9]In the following chapters, we will neglect the notation PD and indicate by D the divisor as well as its dual form. Which of the two interpretations is the appropriate one will always be clear from the context.

Chapter 3

A new offspring of PALP

PALP [1], a package for analyzing lattice polytopes, has the ability to construct, manipulate and analyze lattice polytopes. The present program mori.x, which is also included in the new version of PALP (starting from release 2.0) available at [34], adds further functionalities concerning Calabi-Yau (CY) threefold hypersurfaces.

The main purpose of mori.x is the computation of the Mori cone of toric varieties given by star triangulations of reflexive polytopes, which correspond to crepant subdivisions of the associated fans. The program is able to perform such triangulations for four–dimensional polytopes with up to three non–vertex points if the secondary fan is at most three–dimensional. The program can also be used with a known triangulation as its input starting from PALP release 2.1. This option, which was not contained in PALP 2.0 as described in [35], works for arbitrary dimensions.

3.1 General aspects of mori.x

The program mori.x is part of the latest version of the package PALP that is available at the web page [34] under the GNU license terms. There, the reader can find a detailed installation description. In short, type "make all" in the command line to compile the entire suite of programs. Otherwise, type "make mori" to compile mori.x only. Consult [1] and the more comprehensive [] for the documentation of the other functions of the package: poly.x, cws.x, class.x and nef.x.

We distinguish two types of functionalities of mori.x. The first kind yields information about the appropriately resolved ambient space (see options -g, -I, -m, -P, -K below). This includes the Stanley-Reisner (SR) ideal (with -g) as well as specific information on the geometry of the lattice polytope that determines the ambient toric variety: incidence structure of the facets (-I), IP-simplices (-P) and subdivisions of the fan (-g); furthermore, the Oda-Park algorithm [36, 37] is used to find the Mori cone of the ambient space (-m). The second kind of functionalities deals with the intersection ring (-i, -t) and topological quantities (-b, -c, -d) of the embedded hypersurface. They are determined with the help of SINGULAR [38], a computer algebra system for polynomial computations. Correspondingly, the options -b, -i,

-c, -t, -d (as well as -a, -H, see below) need SINGULAR to be installed.

The generators of the Mori cone are given in terms of their intersections with the toric divisors. For singular toric varieties, the Picard group of Cartier divisors is a non-trivial subgroup of the Chow group, which contains the Weil divisors. Hence one can consider the Kähler cone, which is dual to the Mori cone, as a cone in the vector space spanned by the elements of either the Picard or the Chow group. The program mori.x only deals with simplicial toric varieties, for which the Picard group is always a finite index subgroup of the Chow group [21, 39]. Hence the Cartier divisors are integer multiples of the Weil divisors and this ambiguity does not arise.

Starting with PALP 2.1, mori.x affords two distinct modes of operation. If used with the option -M arbitrary reflexive polytopes of any dimension can serve as input (at least in principle),see section 3.2 for details. Without -M the program only works if the input polytope can be triangulated by mori.x or if it does not require triangulation, as we will outline in the following sections.

As described in [2], mori.x can perform star triangulations of certain four–dimensional reflexive polytopes. This operation was designed for the CY hypersurface case. Generic CY hypersurfaces avoid point-like singularities of the ambient space as well as divisors that correspond to interior points of facets. Consequently, the algorithm performs star triangulations only up to such interior points.

Polytopes can be triangulated by subdividing the secondary fans of its non-simplicial facets [40, 41]. This triangulation algorithm is implemented in mori.x for polytopes with up to three points that are neither vertices nor interior to the polytope or one of its facets; this implies that the secondary fan of any facet can be at most three–dimensional. The program exits with a warning message if the subdivision is not properly completed.

As the dimension of the secondary fan corresponding to a facet grows with the number of points in the facet, this limitation tends to become relevant for toric varieties for which $h^{1,1}$ is large: $h^{1,1}$ increases with the number of points on the polytope and polytopes with many points are more likely to have facets containing many points.

Complete triangulations of arbitrary polytopes can be performed programs such as TOPCOM [42], which is also included in the open source mathematics software system Sage [43]. Sage also contains various tools for handling toric varieties. The triangulations performed in mori.x are attuned to the case of three-dimensional CY hypersurfaces. This means, in particular, that interior points of facets are ignored: one must use -M to avoid this. For small Picard numbers, mori.x is hence faster than programs which perform a complete triangulation.

If a polytope of arbitrary dimension has only simplicial facets whose only lattice points are its vertices and possibly interior points, it does not require any triangulation. Hence mori.x can also handle such cases without -M.

With the option -H the program can also analyze arbitrary hypersurfaces embedded in the ambient toric varieties. It is capable of computing the intersection ring and certain characteristic classes. Here the omission of interior points of facets, which happens as a consequence of mori.x's triangulation algorithm, may introduce severe singularities which often result in non–

integer intersection numbers. There is a warning if there are indeed points interior to facets; in such a case it is probably better to repeat the computation with the combination -HM.

The help screen provides essential information about all the functionalities of the program:

```
palp$ mori.x -h
This is ''mori.x'':
                 star triangulations of a polytope P* in N
                 Mori cone of the corresponding toric ambient spaces
                 intersection rings of embedded (CY) hypersurfaces
Usage:   mori.x [-<Option-string>] [in-file [out-file]]
Options (concatenate any number of them into <Option-string>):
 -h     print this information
 -f     use as filter
 -g     general output: triangulation and Stanley-Reisner ideal
 -I     incidence information of the facets (ignoring IPs of facets)
 -m     Mori generators of the ambient space
 -P     IP-simplices among points of P* (ignoring IPs of facets)
 -K     points of P* in Kreuzer polynomial form
 -b     arithmetic genera and Euler number
 -i     intersection ring
 -c     Chern classes of the (CY) hypersurface
 -t     triple intersection numbers
 -d     topological information on toric divisors &
        del Pezzo conditions
 -a     all of the above except h, f, I and K
 -D     lattice polytope points of P* as input (default CWS)
 -H     arbitrary (also non-CY) hypersurface
        'H = c1*D1 + c2*D2 + ...' input: coefficients 'c1 c2 ...'
 -M     Stanley-Reisner ideal and Mori generators with an
        arbitrary triangulation as input; must be combined with -D
Input:  1) standard: degrees and weights
           'd1 w11 w12 ... d2 w21 w22 ...'
        2) alternative (use -D): 'd np' or 'np d'
           (d=Dimension, np=#[points]) and (after newline) np*d
           coordinates
Output:    as specified by options
```

Following PALP's notation we refer to the M lattice polytope which determines the CY hypersurface as P; consequently its dual, which gives rise to the fan of the ambient toric variety, is P^*.

As PALP always interprets the input as $P \subset M_{\mathbb{R}}$ unless some option modifies this behavior, matrix input of $P^* \subset N_{\mathbb{R}}$ requires the option -D. In order to avoid errors, matrix input is not allowed unless this option is set. If only P but not P^* is known one can use `poly.x -e` to obtain the latter.

3.2 Options of `mori.x`

This section contains a detailed description of the options listed in the help screen. If no flag is specified, the program starts with the parameter `-g`. By default, the program considers a CY hypersurface embedded in the ambient toric variety. The option `-H` has to be used in order to consider non-CY hypersurfaces. Note that the options `-b`, `-i`, `-c`, `-t`, `-d`, `-a`, `-H` need SINGULAR [38] to be installed.

Most options of `mori.x` produce output that is related to the points of P^* in a specific order which can be determined by combining the desired functionality with the option `-P` (see sec. 3.2 below). In order to avoid repeating this information for every option, we now present an example that will be used for many of the options below:

```
palp$ mori.x -P
Degrees and weights  'd1 w11 w12 ... d2 w21 w22 ...':
8 4 1 1 1 1 0  6 3 1 0 1 0 1
4 8  points of P* and IP-simplices
   -1   0   0   0   1   3   1   0
    0   0   0   1   0  -1   0   0
   -1   1   0   0   0   3   1   0
    1   0   1   0   0  -4  -1   0
------------------------------------   #IP-simp=2
    4   1   0   1   1   1   8=d  codim=0
    3   0   1   1   0   1   6=d  codim=1
```

The output above the dashed line just means that P^* has the lattice points[1]

$$p_1 = \begin{pmatrix} -1 \\ 0 \\ -1 \\ 1 \end{pmatrix}, \quad p_2 = \begin{pmatrix} 0 \\ 0 \\ 1 \\ 0 \end{pmatrix}, \quad \ldots, \quad p_7 = \begin{pmatrix} 1 \\ 0 \\ 1 \\ -1 \end{pmatrix}, \quad p_8 = \begin{pmatrix} 0 \\ 0 \\ 0 \\ 0 \end{pmatrix} \quad (3.1)$$

and the last two lines encode the facts $4p_1+p_2+p_4+p_5+p_6 = 0$, $3p_1+p_3+p_4+p_6 = 0$. Note how the dashed line proceeds only up to p_6. This is because `mori.x` always ignores the origin, and, if used without `-M`, also ignores points that are interior to facets: $p_7 = (p_2 + p_4 + p_5 + p_6)/4 = (p_3+p_4+p_6)/3$ lies inside the facet with vertices p_2, p_3, p_4, p_5, p_6. The reader is invited to check that the same example with `mori.x -PM` results in a dashed line below all points except the origin.

---
[1]Here the index starts at 1 instead of 0 as it is standard in PALP. This shift is needed to match the counting of toric divisor classes displayed in certain outputs of `mori.x` and hence avoids confusion.

-h

This option prints the help screen.

-f

This parameter suppresses the prompt of the command line. This is useful if one wants to build pipelines or shorten the input; e.g. our standard example (3.1) can be entered as

```
palp$  echo  '8 4 1 1 1 1 0   6 3 1 0 1 0 1' | mori.x -fP
4 8  points of P* and IP-simplices
...
```

-g

This triggers the general output. First, the triangulation data of the facets is displayed. The number of triangulated simplices is followed by the incidence structure of the simplices. The incidence information for each simplex is encoded in terms of a bit sequence:there is a digit for each relevant polytope point; a 1 denotes that the point belongs to the simplex. Second, the SR ideal is displayed: the number of elements of the ideal is followed by its elements. Each element is denoted by a bit sequence as above.

```
palp$  echo  '8 4 1 1 1 1 0   6 3 1 0 1 0 1' | mori.x -fg
8 Triangulation
110101 111100 101011 101110 100111 111001 001111 011101
2 SR-ideal
010010 101101
9 Triangulation
110101 111100 101011 101110 100111 111001 010111 011011 011110
2 SR-ideal
110010 001101
```

The program performs the two possible triangulations of the facet ⟨23456⟩, which is the only non–simplicial one (see section 3.2). The last two bit sequences of the first result describe the simplices ⟨$\widehat{25}$346⟩, whereas the second triangulation gives the three simplices ⟨$\widehat{2346}$5⟩ (in this notation the hat indicates that one of the points is dropped). Nevertheless, the two resolutions give the same CY intersection polynomial.[2]

---

[2]This fact suggests that the two resolutions give rise to the same CY hypersurface. Indeed, simply connected CY threefolds are completely determined up to diffeomorphisms by their Hodge numbers, intersection rings and second Chern classes [32,33].

-I

The incidence structure of the facets of the polytope P^* is displayed. Interior points of the facets are neglected.

```
palp$ echo '8 4 1 1 1 1 0  6 3 1 0 1 0 1' | mori.x -fI
Incidence: 110101 111100 011111 101011 101110 100111 111001
```

The incidence data show the intersections of $p_1, \ldots, p_6$ (ignoring p_7, p_8!) with the seven facets. The third facet contains the five points $p_2, \ldots, p_6$, hence it is not simplicial and needs to be triangulated.

-m

The Mori cone generators of the ambient space are displayed in the form of a matrix.[3] Each row corresponds to a generator. The entries of each row are the intersections of the generator with the toric divisor classes. The Oda-Park algorithm is used to compute the generators. Furthermore, the incidence structure between the generators of the Mori cone and its facets is displayed. For the standard example this takes the following form.

```
palp$ echo   '8 4 1 1 1 1 0  6 3 1 0 1 0 1' | mori.x -fm
2 MORI GENERATORS / dim(cone)=2
    3  0  1  1  0  1    I:10
    0  3 -4 -1  3 -1    I:01
2 MORI GENERATORS / dim(cone)=2
    1  1 -1  0  1  0    I:10
    0 -3  4  1 -3  1    I:01
```

The Mori cone is two-dimensional, so that its facets can be identified with the generators. This explains the trivial incidence structure.

Let us consider another simple example, a hypersurface in $\mathbb{P}^2 \times \mathbb{P}^1 \times \mathbb{P}^1$.

```
palp$ mori.x -m
Degrees and weights  'd1 w11 w12 ... d2 w21 w22 ...':
3 1 1 1 0 0 0 0  2 0 0 0 1 1 0 0  2 0 0 0 0 0 1 1
3 MORI GENERATORS / dim(cone)=3
    0  0  0  1  1  0  0    I:110
    1  1  0  0  0  0  1    I:101
    0  0  1  0  0  1  0    I:011
```

The Mori cone generators can easily be seen to be dual to the hyperplane sections. Now, the Mori cone is three-dimensional, so that each of its facets contains two generators. Let us, for

---

[3] As divisors corresponding to interior points of facets do not intersect a CY hypersurface, such divisors are neglected in the computation of the Mori cone of the ambient space.

instance, consider the incidence structure between the first generator and the three facets of the Mori cone. Here, the string `I:110` tells that the vector lies on the first and second facets but does not intersect the third one.

For an example with a more complicated structure of the Mori cone see section 3.2.

-P

First a list of lattice points of P^* is displayed in the following manner. If `-P` is combined with both `-M` and `-D`, the list is just the input provided by the user, in the same order except for the fact that the lattice origin comes at the end of the list. In all other cases the complete list of lattice points of P^* is given in the following order:

1. vertices (with `-D` in the order provided by the user),
2. points not interior to the polytope or its facets,
3. points interior to facets,
4. the lattice origin.

Then a dashed line indicates which points are 'relevant': all points except for the origin in the case of `-M`, but not points interior to facets otherwise. Finally the IP-simplices with vertices among these relevant points are displayed.

The output for the standard example can be found above equation (3.1). The following example features all types of lattice points:

```
palp$ echo '16 8 4 2 1 1' | mori.x -fP
4 9  points of P* and IP-simplices
   -1    0    0    2    0    0    0    1    0
   -1    0    0    1    2    0    1    1    0
    0    0    2   -1    1    1    1    0    0
    0    1    1    0   -1    1    0    0    0
--------------------------------------  #IP-simp=3
    8    1    1    4    2    0    0   16=d   codim=0
    4    0    0    2    1    1    0    8=d   codim=1
    2    0    0    1    0    0    1    4=d   codim=2
```

$p_1, \ldots, p_5$ are vertices, p_6, p_7 further relevant points, but $p_8 = -p_1 = (p_2 + p_3 + 4p_4 + 2p_5)/8$ is interior to the facet spanned by $p_2, \ldots, p_5$. Note that the ordering of the CWS input is not obeyed by the output of lattice points. Once the order is displayed, however, it is fixed and determines the labeling of toric divisors in any further output.

-K

The Kreuzer polynomial[4] of PALP's representation of P^* is displayed. It encodes lattice polytope points in a compact form. The number of variables equals the dimension of the polytope. Each lattice point gives rise to a Laurent monomial in which the exponents of the variables are the coordinates. Vertices and non-vertices are distinguished by coefficients '+' and '−' respectively. Points in the interior of facets are ignored. As this is closely connected with the way `mori.x` works when used without -M, the combination -MK is not allowed.

```
palp$ echo  '8 4 1 1 1 0  6 3 1 0 1 0 1' | mori.x -fK
KreuzerPoly=t_4/(t_1t_3)+t_3+t_4+t_2+t_1+t_1^3t_3^3/(t_2t_4^4);
intpts=1;  Pic=2
```

A comparison with the output for -P, which can be found above equation (3.1), might help for a better understanding of the present option.

Negative coordinates are always displayed by putting the variables in the denominator. The number of points in the interior of facets is shown as `intpts`. The multiplicities of the toric divisors are displayed as `multd` if they are greater than one. Furthermore, the Picard number of the CY hypersurface is computed and printed as `Pic`.

-b

The zeroth and first arithmetic genera of the hypersurface are determined according to the following formulas [44]:

$$\chi_q(X) = \sum_p (-1)^p h^{p,q}(X) = \int_X \mathrm{ch}(\Omega^q(X)) \mathrm{Td}(X), \quad q = 0, 1. \tag{3.2}$$

where $\Omega^0(X) = \mathcal{O}_X$ is the trivial bundle and $\Omega^1(X) = T^*X$ is the bundle of 1-forms, ch is the corresponding Chern character, and $\mathrm{Td}(X)$ is the Todd class of X.

Furthermore the Euler characteristic is displayed. Here we compute it by means of the intersection polynomial:

$$\chi = \int_X c_n. \tag{3.3}$$

where c_n is the top Chern class, $n = \dim X$.

These formulas hold for arbitrary smooth hypersurfaces; in particular, they do not need to be CY. Indeed, if X is CY, its Euler characteristic can also be computed by `poly.x` in terms of polytope combinatorics. Compare the two Euler characteristics for a consistency check.

Consider the $K3$ surface as a simple example:

```
palp$ echo '4 1 1 1 1' | mori.x -bf
SINGULAR   -> Arithmetic genera and Euler number of the CY:
chi_0:  2 , chi_1: -20  [ 24 ]
```

---

[4] We named this output format after Maximilian Kreuzer, who designed it. This is an example of his proverbial ability to eliminate unnecessary data redundancies and recast essential information in condensed form.

As expected, the Euler characteristic is 24 and $h^{1,1} = 20$. Using the example discussed before we find

```
palp$ echo '8 4 1 1 1 1 0 6 3 1 0 1 0 1' | mori.x -bf
SINGULAR  -> Arithmetic genera and Euler number of the CY:
chi_0: 0 , chi_1: 126  [ -252 ]
SINGULAR  -> Arithmetic genera and Euler number of the CY:
chi_0: 0 , chi_1: 126  [ -252 ]
```

Special care is needed in the interpretation of the results for non-CY hypersurfaces: the triangulation algorithm might fail to make these varieties smooth, in which case the formulas above do not hold and hence the output is misleading; see the description of the option -H for more details.

-i

This option displays the intersection polynomial in terms of an integral basis of the toric divisors. The coefficients of the monomials are the triple intersection numbers in this basis. This option can also be used together with -H to perform this task for non-CY hypersurfaces.

```
palp$ echo  '8 4 1 1 1 0  6 3 1 0 1 0 1' | mori.x -fi
SINGULAR -> divisor classes (integral basis J1 ... J2):
d1=J1+3*J2, d2=J1, d3=-J1+J2, d4=J2, d5=J1, d6=J2
SINGULAR -> intersection polynomial:
2*J1*J2^2+2*J2^3
SINGULAR -> divisor classes (integral basis J1 ... J2):
d1=J1+3*J2, d2=J1, d3=-J1+J2, d4=J2, d5=J1, d6=J2
SINGULAR -> intersection polynomial:
2*J1*J2^2+2*J2^3
```

d1, ..., d6 denote the toric divisors corresponding to the lattice points $p_1, \ldots, p_6$, cf. eq. (3.1). There are two independent divisor classes. Indeed, mori.x expresses the intersection polynomial in terms of the integral basis $J_1 = D_2 = D_5$ and $J_2 = D_4 = D_6$.

-c

The Chern classes of the hypersurface (CY or non-CY) are displayed in terms of an integral basis of the toric divisors:

```
palp$ echo '8 4 1 1 1 1 0 6 3 1 0 1 0 1' | mori.x -fc
SINGULAR -> divisor classes (integral basis J1 ... J2):
d1=J1+3*J2, d2=J1, d3=-J1+J2, d4=J2, d5=J1, d6=J2
SINGULAR  -> Chern classes of the CY-hypersurface:
```

```
c1(CY)=  0
c2(CY)=  10*J1*J2+12*J2^2
c3(CY)=  -252 *[pt]
SINGULAR -> divisor classes (integral basis J1 ... J2):
d1=J1+3*J2, d2=J1, d3=-J1+J2, d4=J2, d5=J1, d6=J2
SINGULAR  -> Chern classes of the CY-hypersurface:
c1(CY)=  0
c2(CY)=  10*J1*J2+12*J2^2
c3(CY)=  -252 *[pt]
```

-t

The triple intersection numbers of the toric divisors are displayed. The form of this output[5] is designed for further use in Mathematica [45]. Before computing the intersection ring for our standard example (3.1), let us state some expectations. Inspection of the data of the polytope reveals that it describes a K3 fibration with the fiber determined by the weight system 6 3 1 1 1. There are only the two points p_2, p_5 outside the corresponding 3–plane, so each of them must represent the generic fiber with self–intersection 0. In other words, the self–intersections of d_2 and d_5 as well as $d_2 \cdot d_5$ must all vanish. This is confirmed by the following excerpt from the output:

```
echo '8 4 1 1 1 0  6 3 1 0 1 0 1' | mori.x -ft
SINGULAR -> triple intersection numbers:
d6^3->2,
d5*d6^2->2,
d4*d6^2->2,
d3*d6^2->0,
d2*d6^2->2,
d1*d6^2->8,
d5^2*d6->0,
d4*d5*d6->2,
d3*d5*d6->2,
d2*d5*d6->0,
d1*d5*d6->6,
d4^2*d6->2,
d3*d4*d6->0,
d2*d4*d6->2,
d1*d4*d6->8,
d3^2*d6->-2,
d2*d3*d6->2,
```

---

[5]The pre-compiler command DijkEQ in the C file SingularInput.c controls the symbol '->' in option -t.

```
d1*d3*d6->2,
d2^2*d6->0,
d1*d2*d6->6,
d1^2*d6->30,
d5^3->0,
d4*d5^2->0,
d3*d5^2->0,
d2*d5^2->0,
d1*d5^2->0,
d4^2*d5->2,
d3*d4*d5->2,
d2*d4*d5->0,
d1*d4*d5->6,
d3^2*d5->2,
d2*d3*d5->0,
d1*d3*d5->6,
d2^2*d5->0,
d1*d2*d5->0,
d1^2*d5->18,
[...]
```

-d

This option displays topological data of the toric divisors restricted to the (CY or non-CY) hypersurface. The Euler characteristics of the toric divisor classes and their arithmetic genera are shown.

Furthermore, in the case of a three-dimensional hypersurface, the program checks the del Pezzo property against two necessary conditions and analyses the mutual intersections of the del Pezzo candididates. The number of del Pezzo candidates is displayed followed by their type in parenthesis; furthermore, those among them that do not intersect other del Pezzo candidates are listed.

For a del Pezzo divisor S of type n, the following equations should hold:

$$\int_S c_1(S)^2 = 9-n, \quad \int_S c_2(S) = n+3 \implies \chi_0(S) = \int_S \mathrm{Td}(S) = 1. \quad (3.4)$$

Here, Td(S) denotes the Todd class of S, which gives the zeroth arithmetic genus of S upon integration. This test also allows to determine the type of the del Pezzo surface in question. A second necessary condition comes from the fact that a del Pezzo surface is a two-dimensional Fano manifold. Hence, the first Chern class of S integrated over all curves on S has to be positive:

$$D_i \cap S \cap c_1(S) > 0 \quad \forall D_i : D_i \neq S, \quad D_i \cap S \neq 0. \quad (3.5)$$

This condition would be sufficient if we were able to access *all* curves of the hypersurface. In our construction, however, we can only check for curves induced by toric divisors. This functionality was added to carry out the analysis of base manifolds for elliptic fibrations in [46].

Consider the following example: it is well-known that the del Pezzo surface dP_6 can be realized as a homogeneous polynomial of degree 3 in $\mathbb{CP}^3$. Hence a Calabi-Yau hypersurface in a toric variety with CWS

$$5\;1\;1\;1\;1\;1\;0$$
$$2\;0\;0\;0\;0\;1\;1$$

i.e. a $\mathbb{CP}^1$ fibration over $\mathbb{CP}^3$ contains a dP_6: setting the last coordinate z_6 to zero forces all terms to be of the form $z_5^2 P_3(z_1,\ldots,z_4)$, where $P_3(z_1,\ldots,z_4)$ is a homogeneous polynomial of degree 3 in $z_1,\ldots,z_4$. We may set z_5 to 1 by using the second $\mathbb{C}^*$ action, so that the divisor D_6 corresponds to a homogenous polynomial of degree 3 in $\mathbb{CP}^3$, i.e. a dP_6.

This is confirmed by

```
palp$ mori.x -d
Degrees and weights  'd1 w11 w12 ... d2 w21 w22 ...':
5 1 1 1 1 1 0 2 0 0 0 0 1 1
SINGULAR -> topological quantities of the toric divisors:
Euler characteristics: 46 46 46 9 46 55
Arithmetic genera: 4 4 4 1 4 5
dPs: 1 ; d4(6)   nonint: 1 ; d4
```

Note that PALP has exchanged the ordering of the divisors, so that the dP_6 is now given by D_4. This divisor does not intersect any other del Pezzo as it is the only del Pezzo candidate in this example.

-a

This is a shortcut for -gmPbictd.

-D

An alternative way to provide the input is to type lattice polytope points of P^* directly. In this case, one has to use the parameter -D. Let us reconsider the example of sec. 3.2:

```
palp$ mori.x -DP
'#lines #columns' (= 'PolyDim #Points' or '#Points PolyDim'):
4 5
Type the 20 coordinates as dim=4 lines with #pts=5 columns:
-1  2  0  0  0
-1  1  2  0  0
```

```
    0 -1  1  0  2
    0  0 -1  1  1
4 9 points of P* and IP-simplices
   -1  2  0  0  0  0  0   1  0
   -1  1  2  0  0  0  1   1  0
    0 -1  1  0  2  1  1   0  0
    0  0 -1  1  1  1  0   0  0
---------------------------------   #IP-simp=3
    8  4  2  1  1  0  0  16=d  codim=0
    4  2  1  0  0  1  0   8=d  codim=1
    2  1  0  0  0  0  1   4=d  codim=2
```

Note how the order of the vertices corresponds to that of the input (cf. section 3.2).

-H

Using this option, one can specify a (non-CY) hypersurface. The user determines the hypersurface divisor class $H = \sum_i c_i D_i$ in terms of the toric divisor classes D_i by typing its coefficients c_i. The hypersurface can then be analyzed by combining -H with other options, as described above. Just using -H, the program runs -Hb.

The reader is warned: smoothness is not guaranteed anymore, so that the intersection numbers can become fractional. Some choices of the hypersurface equation may intersect singularities not resolved by the triangulation. Consider e.g. the hypersurface determined by the divisor class $H = D_1 + D_6$ in our example (3.1). Remember that the order in which mori.x expects the coefficients of the hypersurface divisor class is fixed by the polytope matrix and not by the CWS input. Hence, the correct input for H is the string 1 0 0 0 0 1.

```
palp$ mori.x -H
Degrees and weights  'd1 w11 w12 ... d2 w21 w22 ...'
8 4 1 1 1 0  6 3 1 0 1 0 1
WARNING: there is 1 facet-IP ignored in the triangulation.
This may lead to unresolved singularities in the hypersurface.
Type the 6 (integer) entries for the hypersurface class:
1 0 0 0 0 1
Hypersurface degrees: ( 5  4 )
Hypersurface class: 1*d1 1*d6
SINGULAR  -> Arithmetic genera and Euler number of H:
chi_0: 29/27 , chi_1: 128/27  [ -22/3 ]
```

To calculate these quantities, the program determines the characteristic classes of the divisors using adjunction. It then performs the appropriate integration with the help of the triple intersection numbers. The fractional results of the arithmetic genera and the Euler number in our example indicate that the intersection polynomial has fractional entries. This happens because the program introduces a singularity into the ambient toric variety which descends to the hypersurface H. It is therefore much better to combine -H with -M:

```
palp$ mori.x -HM
Degrees and weights  'd1 w11 w12 ... d2 w21 w22 ...':
8 4 1 1 1 1 0  6 3 1 0 1 0 1
4 8
  -1   0   0   0   1   3   1   0
   0   0   0   1   0  -1   0   0
  -1   1   0   0   0   3   1   0
   1   0   1   0   0  -4  -1   0
'#triangulations':
1
1 triangulations:
12 1111000 1110010 1101010  0111001 0110011 0101011
1011100 1010110 1001110  0011101 0010111 0001111
Type the 7 (integer) entries for the hypersurface class:
1 0 0 0 0 1 0
Hypersurface degrees: ( 5  4  1 )
Hypersurface class: 1*d1 1*d6
SINGULAR  -> Arithmetic genera and Euler number of H:
chi_0:  1 , chi_1:  3  [ -4 ]
```

As a second example, consider the quadric in $\mathbb{CP}^3$:

```
palp$ mori.x -H
Degrees and weights  'd1 w11 w12 ... d2 w21 w22 ...':
4 1 1 1 1
Type the 4 (integer) entries for the hypersurface class:
2 0 0 0
Hypersurface degrees: ( 2 )
Hypersurface class: 2*d1
SINGULAR  -> Arithmetic genera and Euler number of H:
chi_0:  1 , chi_1: -2  [ 4 ]
```

The hypersurface is smooth in this case, so that the arithmetic genera and the Euler number are those of $\mathbb{CP}^1 \times \mathbb{CP}^1$. Of course, one needs to independently check smoothness in order to rely on the output, as the integrality of the arithmetic genera alone is not sufficient to conclude that the hypersurface is non-singular.

-M

Option -M allows polytopes of (in principle) arbitrary dimensions, but expects the triangulations to be provided by the user; it can be combined with any other option except for -K. As the Mori cone is analysed with PALP's routines, the parameter POLY_Dmax must possibly be adjusted for this.

This functionality is useful, for instance, when mori.x fails to triangulate the polytope by itself. This happens whenever the dimension of the polytope is different from four, or when the polytope contains more than three lattice points that are neither vertices nor (facet–)IPs. Fortunately, there are programs capable of efficiently performing complete triangulations of arbitrary polytopes [42, 43]; the user can redirect their output as an input for mori.x to determine the Mori generators of the ambient space. Other situations where -M is useful arise whenever we prefer to keep control over the lattice points involved in the triangulation, rather than accept mori.x's convention of omitting precisely the interior points of facets from a completed list; this is particularly relevant if -M is combined with -H.

After the usual polytope input, P^* is displayed in the following manner. If the input is of CWS type, all points of P^* are given in mori.x's standard order (see section 3.2). If matrix input is used via -D, the points entered by the user are displayed again in the same order, but with the origin appended (if the origin is accidentally entered somewhere in the point list, it is swapped with the last point in the list); possible further polytope points are ignored by the program, hence singularities can be introduced if desired. Then the user is asked for the number of triangulations to be analysed, and afterwards each triangulation should be entered as a line starting with the number of simplices involved in the triangulation, followed by bit sequences encoding these simplices. The number of bits in each sequence should be the number of non–zero lattice points in the displayed list, with 1's indicating that the point belongs to the simplex, and 0's otherwise.

An application to our standard example (3.1) was already demonstrated in section 3.2. Consider also the following two-dimensional polytope:

```
palp$ mori.x -MDgm
'#lines #columns' (= 'PolyDim #Points' or '#Points PolyDim'):
2 7
Type the 14 coordinates as dim=2 lines with #pts=7 columns:
    1    0   -1   -1   -1   -1    0
    0    1    2    1    0   -1   -1
2 8
    1    0   -1   -1   -1   -1    0    0
    0    1    2    1    0   -1   -1    0
'#triangulations':
1
1 triangulations:
7 1100000 0110000 0011000 0001100 0000110 0000011 1000001
```

```
14 SR-ideal
0000101 0001001 0001010 0010001 0010010 0010100 0100001 0100010
0100100 0101000 1000010 1000100 1001000 1010000
6 MORI GENERATORS / dim(cone)=5
   1 -2  1  0  0  0  0   I:0111011
   0  1 -1  1  0  0  0   I:1101101
   0  0  1 -2  1  0  0   I:1110110
   0  0  0  1 -2  1  0   I:1011111
   0  0  0  0  1 -1  1   I:1100011
   1  0  0  0  0  1 -1   I:0111100
```

Since P^* is just a polygon there is only one maximal triangulation of its boundary. For this reason we have typed **1** after '**#triangulations**':. The triangulation has seven simplices, which are just the line segments along the circumference of the polygon. For instance, the string **0011000** denotes the third simplex containing the points denoted by the third and fourth columns of the polytope matrix, i.e. the points $p_3 = (-1, 2)$ and $p_4 = (-1, 1)$.

As the Mori cone encodes linear relations among seven points in $d = 2$, it is five-dimensional and has four-dimensional facets; there are seven of them, as the lengths of the bit sequences after **I:** indicate. There are six generators. Consider the matrix of incidences whose rows are preceded by **I:**. The second column reads **111011**, i.e. on the second facet of the Mori cone lie five generators. It is easily checked that they satisfy $m_1 + 2m_2 + m_3 = m_5 + m_6$. All other facets contain instead four generators and are hence simplicial.

3.3 Structure of the program

mori.x is part of the new releases of PALP (starting from version 2.0). The general structure of the package has not been changed, only some new files have been added. Hence, all general annotations to the package in [1] remain valid. In this section, we provide an overview of the composition of **mori.x** and discuss its dependencies on pre-existing files.

The source code of **mori.x** is contained in the program files *mori.c*, *MoriCone.c*, *SingularInput.c*, and the header file *Mori.h*. *Makefile* reflects the dependencies of the program. The compilation supports 32 as well as 64 bit architectures. One can adjust the compilation parameters according to ones needs. The optimization level is set at **-O3** by default.

mori.c contains the main and the help information routines. Further, basic manipulations (completion, calculation of facet equations,...) of the polyhedron are performed with the help of core routines from *Vertex.c*. In particular, reflexivity of the polytope is checked.

MoriCone.c is at the heart of **mori.x**. After determining the non-simplicial facets, their triangulation is performed by the routine **GKZSubdivide**. This function identifies the maximal dimensional secondary fans of the facets and makes a case-by-case triangulation depending on their dimensions. This function is only implemented for secondary fans up to dimension three.[6] Once the subdivision is accomplished, the program determines the Stanley-Reisner ideal (**StanleyReisner**) and computes the Mori cone (**Print_Mori**). Furthermore, it finds a basis of the toric divisor classes.

SingularInput.c is the interface to SINGULAR [38]. The latter is a very efficient computer algebra system for computations with polynomial rings. In *SingularInput.c*, the Chow ring is determined from the Stanley-Reisner ideal, the linear relations among the toric divisors, and a basis of the toric divisors of the triangulated polytope. This data is put together by **HyperSurfSingular** and redirected to SINGULAR, which then determines the intersection ring restricted to the hypersurface and computes its characteristic classes.[7]

The most important routines of **mori.x** are documented in the file *Mori.h*. This header file provides a more detailed description of the structure of the program.

---

[6]The pre-compiler command TRACE_TRIANGULATION in *MoriCone.c* enables diagnostic information about the triangulation. This data might be of use for the motivated programmer who wants to extend the subdivision algorithm.

[7]The input for SINGULAR can be displayed in the standard output of **mori.x** by turning on the pre-compiler definition TEST_PRINT_SINGULAR_IO in *SingularInput.c*.

Chapter 4

Four-modulus 'Swiss cheese' chiral models

The 'Large Volume Scenario' (LVS), developed in [3], is a new strategy for stabilizing the Kähler moduli in type IIB Calabi-Yau orientifold compactifications. This strategy can be seen as a cousin of the KKLT strategy [47]. In both cases, one first stabilizes the axio-dilaton and complex structure moduli by means of the flux induced Gukov-Vafa-Witten (GVW) superpotential, and then one tries to stabilize the Kähler moduli by non-perturbative effects such as E3-branes (Euclidean D3-branes), and gaugino condensation. The key difference between these two strategies lies in the fact that the LVS admits non-supersymmetric anti-de Sitter minima, whereby the Calabi-Yau volume is exponentially large with respect to the size of the E3-brane, and, at fixed g_s, it is independent of the flux superpotential W_0. This latter fact implies that this non-perturbative stabilization of the Kähler moduli will not mess up the complex structure stabilization. Other advantages of this scenario are explained in [12].

The key requirement to construct an LVS model, is to find a Calabi-Yau threefold with $h^{2,1} > h^{1,1} > 1$, and such that the volume of the manifold is driven by the volume of a single 'large' four-cycle, and that the rest of the four-cycles contribute negatively to the overall volume. This structure has been dubbed the 'Swiss cheese' structure. Because it is possible to make cycles small while keeping the CY large, we can have E3-instantons that make large contributions and have a large volume vacuum. These instanton effects now becoming important, actually compete against α'-corrections to the Kähler potential. Having these 'small', shrinkable cycles also serves another useful purpose. If one places MSSM-like stacks of D7-branes on them, by going to this large volume limit where these are made small, one effectively decouples the gauge theory on the brane from the UV dynamics encoded by the rest of the Calabi-Yau data. In this way, one addresses the comment in [48], which points out a drawback of generic models: namely, that making the volume of the CY large will typically force one to scale up the cycles on which branes are wrapped.

In [4], Blumenhagen et al have shown that the standard two-step model building paradigm, where one first stabilizes the closed string moduli and then introduces MSSM-like D7-branes, is too naïve. Such D-branes would intersect the E3-branes used in the non-perturbative stabilization, thereby inducing charged zero-modes. In order for the E3 contribution to the superpotential to be non-vanishing, one would then have to turn on vev's for charged superfields, thereby spontaneously breaking the MSSM-like gauge symmetry. In that article, a solution to the problem is outlined and explicitly worked out for a three-modulus Calabi-Yau, whereby two intersecting stacks of MSSM-like D7-branes are setup so as not to chirally intersect the E3-brane.

In this chapter, we will address the issue of chiral zero-modes while taking even more stringent constraints into account. Namely, we will take into account the fact that the MSSM-like D7-branes are wrapped on non-spin manifolds, thereby inducing half-integral world-volume fluxes which themselves induce unwanted, charged zero-modes. These fluxes compensate for the open string world-sheet anomaly discovered in [49].

We will *not* attempt to construct realistic MSSM configurations. Our goal will be to have setups with two intersecting D7-brane stacks with unitary gauge groups and bifundamental chiral fermions, that accommodate the LVS scenario of moduli stabilization. These setups can then in principle be used to create inflationary models. We will see that requiring zero chiral intersections between the E3-brane and the MSSM-like branes, and between the 'hidden' D7-branes (that are needed to saturate the negative D7-charge from the O7-planes) and the rest of the branes, will impose heavy restrictions that will rule out some models.

In order to accommodate two MSSM-like D7-stacks and an E3-brane, all on different 'small' cycles, we need Calabi-Yau manifolds with at least four moduli, whereby three preferably come from blow-ups. For this purpose, we will scan through the list of CY hypersurfaces of toric fourfolds encoded as four-dimensional polytopes in [30]. From this list, we will select all four-modulus CYs of which the polytopes have five vertices, which is the minimal amount of possible vertices for four-dimensional polytopes. This will ensure that three of the four moduli will correspond to divisors that originate from blow-ups. We will then proceed to triangulate all relevant polytopes by using a recently enhanced version of the PALP package [1,2]. Furthermore, PALP computes the triple intersection numbers of our CYs and determines the Mori cones of their ambient spaces. After eliminating models with equivalent triangulations, we end up with four 'Swiss cheese' models.

We will see, however, that not all four-cycles that contribute negatively to the overall CY volume can be shrunk arbitrarily while preserving the large volume limit. This will corroborate the analysis of [50] that gives precise conditions for this to be possible. We will also study the topologies of our four-cycles in detail, and will see that not all rigid, 'small', cycles with $h^{0,1} = h^{0,2} = 0$ are del Pezzo surfaces, which is a necessary condition for 'shrinkability'.

This chapter is organized as follows: in section 4.1, we briefly review some definitions relevant to $\mathcal{N} = 1$ type IIB orientifold compactifications, we also review the LVS, and we reiterate the chiral zero-mode issue raised in [4]. In section 4.2, we review how the Freed-Witten anomaly induces half-integral flux when a D-brane wraps a non-spin cycle. In section 4.3, we explain how we count both neutral and charged zero-modes of E3-instantons. Section 4.4 contains our first model. Here, we will be very explicit about our strategy. We will present the toric data, explain how we search for and classify 'small' divisors, and then move on to model building. This will be done in a three-step procedure: first we build 'local' models containing MSSM-like branes and E3-branes without canceling the D7-tadpole. Then we pick an orientifold involution and add 'hidden' branes appropriately so as not to intersect the visible sector. Finally, we study the Kähler moduli stabilization. This model will only be 'half' successful, in the sense that we will be able to solve the chiral intersection problem but will not find a large volume minimum. In section 4.5 we present our second model, which will be successful in this sense, although one unstabilized modulus will remain. In appendix A.1 we present the relevant definitions for B-branes such as induced charges, orientifolding, D-terms and constructions of involution invariant D7-branes. Finally, in appendices A.2 and A.3 we present our two remaining models. All of our results are summarized in table 4.6, in the conclusions.

4.1 Large volume scenario

4.1.1 General idea

We briefly review some definitions for $\mathcal{N} = 1$ flux compactifications of type IIB in order to set our conventions. The full superpotential for type IIB compactified on a CY threefold X is given by

$$W = \int_X G_3 \wedge \Omega_3 + \sum_{i=1}^{h^{1,1}} A_i\left(S, U\right) e^{-a_i T_i}, \tag{4.1}$$

where the first term is the GVW potential [51], which stabilizes the complex structure moduli and the axio-dilaton field, $S = e^{-\phi} + iC_0$.[1] The second term takes into account non-perturbative corrections to the superpotential. We focus here on corrections due to the presence of E3-brane instantons, in which case the functions A_i only depend on the axio-dilaton and the complex structure moduli.[2] Furthermore, T_i are the Kähler moduli of type IIB orientifolds:

$$T_i = e^{-\phi}\tau_i + i\rho_i. \tag{4.2}$$

Here, τ_i denotes the volume of the divisor D_i, and ρ_i is the corresponding axion field originating from the R-R four-form:

$$\tau_i = \frac{1}{2}\int_{D_i} J \wedge J = \frac{1}{2}\kappa_{ijk}t^j t^k, \quad \text{and} \quad \rho_i = \int_{D_i} C_4. \tag{4.3}$$

The κ_{ijk} coefficients determine the triple intersection numbers given a basis of integral two-forms $\{\eta_i\} \in H^{1,1}(X, \mathbb{Z})$, in which we will choose to expand the Kähler form:

$$J = \sum_i t_i \eta_i. \tag{4.4}$$

---

[1]In this chapter, we use a different notation than that described in chapter 1. Indeed, to aid comparison, we adopt here the notation of [4]. In particular, note that the definition of the axio-dilaton (here S) differs from (1.6).

[2]In general non-perturbative corrections can also arise from gaugino condensation from wrapped D7-branes.

The Kähler potential with its leading α'-correction [52] takes the following form:

$$K = -2\ln\left(\hat{\mathcal{V}} + \frac{\xi}{2g_s^{3/2}}\right) - \ln\left(S + \bar{S}\right) - \ln\left(-i\int_X \Omega \wedge \bar{\Omega}\right). \tag{4.5}$$

Where $\xi = -\frac{\zeta(3)\chi(X)}{16\pi^3}$ encodes the perturbative α'-correction in terms of the Euler characteristic of X. The symbol $\hat{\mathcal{V}}$ denotes the volume of the CY in the Einstein frame, where the metric is expressed in terms of the string frame metric by $g^E_{\mu\nu} = e^{-\phi/2}g_{\mu\nu}$. The volume in the string frame is given by

$$\mathcal{V} = \frac{1}{3!}\int_X J \wedge J \wedge J = \frac{1}{6}\kappa_{ijk}t^i t^j t^k. \tag{4.6}$$

Note that in computing the volume of the CY we assume that NS-NS fluxes have stabilized the background value of the dilaton. Hence, we may effectively treat the latter as a constant, and readily switch frames. Strictly speaking, this is only a large volume approximation, as the dilaton will vary strongly in the vicinity of the D7-brane and the O7-plane; see section 1.3. Since we will mainly work in the string frame, from now on we will explicitly denote quantities in the Einstein frame by a hat symbol.

The four-dimensional scalar potential for all moduli fields gets contributions from both F- and D-term potentials. The F-term has the following form

$$V_F = e^K\left(\sum_{i=T,S,U} K^{i\bar{j}}D_i W D_{\bar{j}}\bar{W} - 3|W|^2\right), \tag{4.7}$$

where the sum runs respectively over the Kähler structure, the axio-dilaton and the complex structure moduli. The non-perturbative term in the superpotential depends explicitly on the Kähler moduli T_i, and thus breaks the no-scale structure of the superpotential.

We are interested in CY manifolds characterized by a volume function of the following shape:

$$\mathcal{V} \sim \tau_l^{\frac{3}{2}} - \sum_{s=1}^{h^{1,1}-1} \tau_s^{\frac{3}{2}}. \tag{4.8}$$

The important property of this function lies in the fact that there is one four-cycle that contributes positively to the volume, and the remaining three contribute negatively. This means that, in principle, one can take a limit where the positively contributing cycle is taken large, and the other three are sent small, while keeping the overall volume of the CY large. Hence, the cycle with volume τ_l will be referred to as a 'large' cycle, and the remaining ones as 'small' cycles. For this reason these manifolds are colloquially referred to as 'Swiss cheese' CY manifolds.

The reason why one would like to have such a CY is that it allows for the LVS [3], which we will now briefly describe. Inserting (4.1) and (4.5) in the above formula for the F-term (4.7), the potential splits into three parts: two non-perturbative terms depend explicitly on the Kähler moduli, and one term accounts for the α'-corrections: $V_F = V_{np1} + V_{np2} + V_{\alpha'}$. In the large volume regime these terms behave like

$$V_{np1} \sim \frac{1}{\hat{\mathcal{V}}} a_s^2 |A_s|^2 \left(-\kappa_{ssj} t^j\right) e^{-2a_s\hat{\tau}_s} e^{K_{cs}} + \mathcal{O}\left(\frac{e^{-2a_s\hat{\tau}_s}}{\hat{\mathcal{V}}^2}\right), \tag{4.9}$$

$$V_{np2} \sim -\frac{a_s\hat{\tau}_s e^{-a_s\hat{\tau}_s}}{\hat{\mathcal{V}}^2} |A_s W_0| e^{K_{cs}} + \mathcal{O}\left(\frac{e^{-a_s\hat{\tau}_s}}{\hat{\mathcal{V}}^3}\right), \tag{4.10}$$

$$V_{\alpha'} \sim \frac{3\hat{\xi}}{16\hat{\mathcal{V}}^3} |W_0|^2 e^{K_{cs}} + \mathcal{O}\left(\frac{1}{\hat{\mathcal{V}}^4}\right), \tag{4.11}$$

where $\hat{\xi} = e^{-\frac{3\phi}{2}}\xi$. V_{np1} is positive and proportional to self-intersection of the small cycle. Since we require $h^{2,1} > h^{1,1}$, also $V_{\alpha'}$ contributes positively to the potential. The second term, instead, contributes negatively. If we consider the decompactification limit, maintaining $a_s\hat{\tau}_s = \ln\hat{\mathcal{V}}$, the three terms become proportional to the inverse third power of the CY volume, thus they are all on equal footing. At this point, the potential is negative. But for increasing $\hat{\mathcal{V}}$, V_{np2} grows faster than $V_{np1} + V_{\alpha'}$. Due to the positive contribution of V_{np1} and $V_{\alpha'}$ the potential starts positive by small volume values, then reaches a negative minimum and afterwards approaches zero from below for asymptotically large values of the volume. This ensures the existence of a local anti-de Sitter minimum at finite volume. The 'Swiss cheese' shape of the manifold is needed here to keep the cycle $\hat{\tau}_s$ logarithmically small compared to the overall volume.

Assuming that the complex structure moduli and the axio-dilaton have been stabilized via the GVW superpotential, we can rewrite the F-term potential for the Kähler moduli in the large volume limit following [4, 50, 53, 54]:

$$V_F = \frac{1}{\hat{\mathcal{V}}^2}\left(-4\pi^2 \text{Vol}\left(D_{E3} \cap D_{E3}\right)\hat{\mathcal{V}}|A_{E3}|^2 e^{-4\pi\hat{\tau}_{E3}} \right. $$
$$\left. -4\pi\hat{\tau}_{E3} e^{-2\pi\hat{\tau}_{E3}} |A_{E3}W_0| + \frac{3}{4}\frac{\hat{\xi}}{\hat{\mathcal{V}}}|W_0|^2\right). \tag{4.12}$$

Let us now discuss D-terms. The several D7-branes wrapped on divisors D_i give rise to the following D-term:

$$V_D = \sum_{i=1}^{N} \frac{1}{\text{Re}(f_i)} \left(\sum_j Q_j^{(i)} |\phi_j|^2 - \hat{\xi}_i\right)^2, \tag{4.13}$$

where the ϕ_i are chiral fields charged under the gauge symmetries of the D7-branes. This potential is determined by the real part of the gauge kinetic functions,

$$\text{Re}(f_i) = e^{-\phi} \frac{1}{2} \int_{D_i} J \wedge J - e^{-\phi} \int_{D_i} \text{ch}_2\left(\mathcal{L}_i - B\right) = \hat{\tau}_i - e^{-\phi} c_i; \tag{4.14}$$

and the Fayet-Iliopoulos terms

$$\hat{\xi}_i = -\text{Im}\left(\frac{1}{\hat{\mathcal{V}}} \int_X e^{-(B+i J)} \Gamma_i\right). \tag{4.15}$$

Here Γ_i denote the charge vectors of the D7-branes. See appendix A.1.1 for a definition thereof.

4.1.2 Incorporation of D7-brane stacks

In order to combine the closed string moduli stabilization with a string theoretic realization of the MSSM, the standard paradigm in type IIB string theory describes the gauge groups as arising from D-brane stacks, and the chiral matter from intersections between stacks. This necessarily requires the incorporation of D7-branes, and therefore O7-planes.

As was explained in [4], the standard strategy of first stabilizing all closed string moduli and then adding MSSM-like D7-brane stacks has a serious pitfall. As we will explain in the next section, the D7-branes will in general be forcefully magnetized. Since they will generically intersect the E3-branes, the E3-D7-strings will correspond to chiral zero-modes of the instanton that are charged under the MSSM-like gauge groups. Therefore, in order to saturate the instanton path integral, any non-zero contribution will have to be accompanied by a multiplicative factor of charged superfields. Since we want our LVS models to serve as a first step in creating models that describe the inflationary epoch, during which energies were above the electro-weak breaking scale, we want to keep the MSSM-like gauge group unbroken. This means that charged superfields must have zero vev's, which will then force such charged superpotentials to vanish.

The strategy is then to engineer our models as follows: we will have one E3-brane placed on a 'small' four-cycle, and two MSSM-like D7-branes with unitary gauge groups placed on the two remaining four-cycles. Finally, to cancel the total D7-tadpole by the O7-plane, we need a 'hidden' D7-brane. We will impose the following constraints on the chiral intersections between the branes:

1. Both MSSM-like D7-branes have no net chiral zero-modes with the E3-brane, but do have chiral matter amongst themselves.

2. The 'hidden' D7-brane has no chiral intersections with either the MSSM-like branes, nor the E3-brane.

In the next section, we explain more clearly why D7-branes are forcefully magnetized.

4.2 Freed-Witten anomaly

In order for the open string world-sheet theory to be consistent, the submanifold on which a D-brane is wrapped must be chosen with care. In [49], Freed and Witten worked out two types of pathologies that can arise.

If a D-brane is wrapped on a submanifold W, such that the pullback of the NS-NS three-form field-strength onto W is non-trivial, i.e. $i^*(H) \neq dB$, then the open world-sheet theory has a fatal anomaly that can only be compensated by having lower brane world-volumes end on W. We will not thoroughly analyze this issue in this work, but will make remarks about it whenever possible.

The other possible pathology has to do with the topology of the submanifold W itself. If W does not admit a spin structure, this leads to a world-sheet anomaly, unless one compensates this by 'twisting' the would-be spin bundle with a would-be $U(1)$-bundle (see [55] for a pedagogical explanation of this). Pragmatically, this means that one has to turn on a 'half-integral' Born-Infeld flux equal to $F = -c_1(N_W)/2$. In general, the total flux on a D-brane will be of the form

$$F = -\frac{c_1(N_W)}{2} + \Delta F, \tag{4.16}$$

where $\Delta F \in H^2(W, \mathbb{Z})$. Although this half-integral shift is in some sense artificial, it must be taken seriously for all practical purposes: it will induce lower brane charges and will contribute to the chiral intersections (A.7), as explained in [56]. This latter fact severely constrains the possibility of generating neutral superpotentials, i.e. superpotentials arising from E3-instantons that have no chiral intersections with the D7-branes in the setup.

Although we will not in general be able to determine, whether or not a submanifold is spin, we will at least be able to test, whether the 'visible' effect of the half-integer shift can be canceled by a *bona fide* integral flux in $H^2(W, \mathbb{Z})$. More precisely, we will establish a necessary criterion to test for this possibility.

The formula (A.7) for the chiral intersection between two branes depends on the Born-Infeld fluxes only through the charges they induce. For branes wrapping four dimensional submanifolds, the Dirac-Schwinger-Zwanziger (DSZ) product depends on the D7- and induced D5-charges seen in the total charge vector (A.5). Therefore, the only way the half-integer shift in the flux can do harm, is through the part that survives the push-forward operation

$$F \mapsto \left(\int_W F \cdot \imath^*(D_A) \right) \tilde{D}^A. \tag{4.17}$$

Now suppose there is a two-form $\gamma \in H^2(W, \mathbb{Z})$ such that

$$\int_W \left(-\frac{c_1(N_W)}{2} + \gamma \right) \cdot \imath^*(D_A) = 0 \quad \forall \ D_A \in H^2(X, \mathbb{Z}). \tag{4.18}$$

Clearly, γ can not be a pulled-back form $\gamma \neq \imath^*()$, since it would have to emanate from a half-integral form in X. It could, however, be a two-form that can be decomposed into a pulled-back part in $H^2(X, \mathbb{Q})$ and a part orthogonal to this. Such forms are referred to as 'gluing vectors', (see [57] for definitions). Be that as it may, the two-form γ, which is assumed to be of type $(1, 1)$, must be Poincaré dual to some linear combination of holomorphic curves on W

$$\gamma = \sum_i n_i [C_i], \quad [C_i] \in H^2(W, \mathbb{Z}), \quad n_i \in \mathbb{Z}. \tag{4.19}$$

By virtue of the fact that W is holomorphically embedded in X, and that these curves are holomorphically embedded in W, the latter are also holomorphically embedded in X. Therefore, the induced D5-charges can be written as follows:

$$\begin{aligned} q_{D5,A} \equiv \int_W \gamma \cdot \imath^*(D_A) &= \sum_i n_i \int_{C_i} \imath^*(D_A), \\ &= \sum_i n_i \int_{\imath_*(C_i)} D_A, \end{aligned} \tag{4.20}$$

where, in the last line, we integrate the D_A over the push-forwards $\imath^*(C_i)$ of the curves. Since the latter are well-defined classes in $H_2(X,\mathbb{Z})$, these D5-charges must be integers.

In conclusion, we will apply the following rule: A D7-brane wrapped on a divisor W will at least carry a half-integer flux $F = -c_1(N_W)/2$. If the induced D5-charges are not all integers, then this half-integral shift can not be compensated by turning on a *non* pulled-back flux. If they are all integers, then more information is needed to decide.

4.3 Instanton zero-mode counting

The spacetime effects of D-instantons and their zero-modes can be described by means of CFT. Detailed accounts of this topic in the IIA setting can be found in [58–60], and in the type IIB setting in [61].

4.3.1 Neutral zero-modes

We are interested in finding E3-branes that will induce a four-dimensional non-perturbative superpotential depending on the complexified Kähler modulus corresponding to the divisor of the E3-brane. Witten's well-known criterion for determining whether a specific E3-brane may or may not contribute requires finding an explicit F-theory lift of the type IIB setup. One can also work directly in type IIB and count the number and type of fermionic zero-modes associated with the E3-brane.

In order for the E3-brane to generate a superpotential as opposed to a higher F-term or a D-term, it can not have more than two fermionic neutral zero-modes. Neutral zero-modes arise from strings with both end points on the E3. These modes can be classified into the following three categories:

1. Universal zero-modes: these strings correspond to the four real scalar fields on the worldvolume theory of the E3 parametrizing transverse motion in four-dimensional spacetime, and their four fermionic superpartners. These modes are model-independent, as the name suggests. The integration over these modes must be saturated by operator insertions that will destroy the superpotential structure of the instanton contribution, thereby turning it into a D-term. There are several known mechanisms to get around this issue. One of them is to let the orientifold projection get rid of half of these fermionic zero-modes [62–65]. This requires the E3-brane to be transversally invariant under the orientifold involution, i.e. that it be mapped to itself as a set. Other mechanisms are known (e.g. [66]), but we will focus on the orientifold mechanism.

2. Internal motion of the E3: the divisor D on which the E3 is wrapped can have moduli, which will also correspond to scalar fields in the world-volume field theory. The number of these moduli is given by the number of non-trivial sections of the normal bundle of D:

$$\# \text{ sections } = H^0(D, ND). \tag{4.21}$$

By Serre duality, (or very roughly, by contracting with the holomorphic three-form of X), this dimension is equal to the the Hodge number $h^{0,2}(D)$:

$$\dim H^2(D, \mathcal{O}) = h^{0,2}(D). \tag{4.22}$$

3. Wilson lines: If the divisor D has $h^{0,1} \neq 0$, then the world-volume gauge theory has Wilson line moduli. These can be counted as follows: first, we compute $h^{0,2}$ by counting the number of non-trivial sections of the normal bundle of D. Then, we compute the holomorphic Euler characteristic χ_0 of D:

$$\chi_0 = \int_D \mathrm{Td}(D) = \frac{1}{12} \int_D \left(c_1(D)^2 + c_2(D)\right) = \frac{1}{12} \int_X \left(2D^3 + c_2(X) \wedge D\right). \tag{4.23}$$

From this, we can deduce $h^{0,1} = 1 + h^{0,2} - \chi_0$.

Imposing $h^{0,2} = h^{0,1} = 0$ is a sufficient criterion for the instanton to contribute to a superpotential. However, the latter may be a charged superpotential, as we will see next.

4.3.2 Charged zero-modes

In [4], a very important issue has been raised concerning the generation of an *uncharged* superpotential. If an E3-brane intersects a D7-brane, the strings stretched between them give rise to bifundamental zero-modes that also need to be soaked up. This requires inserting charged chiral superfields Φ_i in the path integral, thereby spoiling the generation of an uncharged superpotential, and leading to something of the form

$$W \sim \prod_i \Phi_i \, e^{-T_{E3}}. \tag{4.24}$$

In order for such a term to be non-zero, one must then require that the charged superfields have vev's; this would induce a breaking of the MSSM-like gauge group. The main point of [4] is that one does not want to break the gauge group at the high energy scale of this setup. Hence, in order to generate phenomenologically viable (uncharged) superpotentials, we must require that the E3 does not intersect any other brane present:

$$\langle \Gamma_{E3}, \Gamma_{D7} \rangle = 0. \tag{4.25}$$

Searching for setups that satisfy this equation will be the main concern of this chapter. The fact that D-branes generically have a half-integral flux that can not be turned off, as explained in the previous section, will severely restrict the possibility of having setups with several D7-brane stacks with none of them intersecting the E3-branes.

4.4 First model

For the sake of clarity, we will give a very detailed account of this first model. We will be more concise in the subsequent models. For a brief introduction to the geometrical methods we used, see chapter 2 and the references therein.

	x_1	x_2	x_3	x_4	x_5	x_6	x_7	x_8	p
	15	10	2	2	1	0	0	0	30
	9	6	1	1	0	1	0	0	18
	7	5	1	1	0	0	1	0	15
	3	2	0	0	0	0	0	1	6

Table 4.1: Projective weights under the toric $\mathbb{C}^*$ actions for the resolved $\mathbb{P}^4_{15,10,2,2,1}(30)$ space. The peculiar order of the coordinates is due to PALP's internal computational optimization.

4.4.1 The resolved $\mathbb{P}^4_{15,10,2,2,1}(30)$ geometry

Toric data

Our first model will be the degree 30 hypersurface of the weighted projective space $\mathbb{P}^4_{15,10,2,2,1}$. Smoothing out this model requires three toric blow-ups, thereby endowing the CY manifold with four Kähler moduli. Table 4.1 shows the homogeneous coordinates of the ambient fourfold and their projective weights under the four $\mathbb{C}^*$ actions.

For the unique triangulation the Stanley-Reisner ideal reads

$$I_{SR} = \{x_1\,x_5,\ x_5\,x_8,\ x_7\,x_8,\ x_1\,x_2\,x_6,\ x_1\,x_2\,x_8,\ x_3\,x_4\,x_5,\ x_3\,x_4\,x_6,\ x_3\,x_4\,x_7,\ x_2\,x_6\,x_7\}. \quad (4.26)$$

In this notation, the entries are coordinates that are not allowed to vanish simultaneously. For instance, the last entry means that x_2, x_6 and x_7 can not vanish simultaneously. The triple intersection numbers of divisor classes[3] in the basis $\eta_1 = D_5$, $\eta_2 = D_6$, $\eta_3 = D_7$, $\eta_4 = D_8$ are encoded in the following polynomial

$$\begin{aligned}I_3 &= 8\eta_1^3 + 8\eta_2^3 - 96\eta_3^3 + 9\eta_4^3 + 3\eta_1^2\eta_2 - 21\eta_1^2\eta_3 \\ &\quad - 5\eta_1\eta_2^2 + \eta_2^2\eta_4 - 3\eta_2\eta_4^2 + 45\eta_1\eta_3^2\,.\end{aligned} \quad (4.27)$$

The Kähler form in the basis $\{\eta_1, \eta_2, \eta_3, \eta_4\}$ is given by

$$J = t_1\eta_1 + t_2\eta_2 + t_3\eta_3 + t_4\eta_4. \quad (4.28)$$

The volumes of the corresponding divisors are

$$\begin{aligned}\tau_1 &= \frac{1}{10}\left((15\,t_3 - 7\,t_1)^2 - (3\,t_2 - 5\,t_2)^2\right), \\ \tau_2 &= \frac{1}{6}\left((3\,t_1 - 5\,t_2)^2 - (t_2 - 3\,t_4)^2\right), \\ \tau_3 &= \frac{1}{14}\left(3\,t_3^2 - 3\,(15\,t_3 - 7\,t_1)^2\right), \\ \tau_4 &= \frac{1}{2}\,(t_2 - 3t_4)^2,\end{aligned} \quad (4.29)$$

---

[3]Throughout this chapter, we will use the sloppy notation where η_i can denote a two-form, a second cohomology class, a divisor, and a line bundle whose first Chern class is given by the denoted two-form. It should, however, always be clear from the context which interpretation is appropriate.

and the volume of the CY manifold is given by

$$\mathcal{V} = \frac{1}{630}\left[45t_3^3 - 3\left(15\,t_3 - 7\,t_1\right)^3 - 7\left(3t_1 - 5t_2\right)^3 - 35\left(t_2 - 3t_4\right)^3\right] \tag{4.30}$$

$$= \frac{\sqrt{2}}{3}\left[\frac{1}{7\sqrt{3}}\left(15\tau_1 + 9\tau_2 + 7\tau_3 + 3\tau_4\right)^{\frac{3}{2}} - \frac{1}{35}\left(5\tau_1 + 3\tau_2 + \tau_4\right)^{\frac{3}{2}} - \frac{1}{15}\left(3\tau_2 + \tau_4\right)^{\frac{3}{2}} - \frac{1}{3}\tau_4^{\frac{3}{2}}\right].$$

It has the expected Swiss cheese form. From this volume formula we deduce the diagonal basis to be

$$\begin{aligned}
D_a &= 15\eta_1 + 9\eta_2 + 7\eta_3 + 3\eta_4\,,\\
D_b &= 5\eta_1 + 3\eta_2 + \eta_4\,,\\
D_c &= 3\eta_2 + \eta_4\,,\\
D_d &= \eta_4\,.
\end{aligned} \tag{4.31}$$

In this basis the total volume reads

$$\mathcal{V} = \frac{\sqrt{2}}{3}\left(\frac{1}{7\sqrt{3}}\tau_a^{\frac{3}{2}} - \frac{1}{35}\tau_b^{\frac{3}{2}} - \frac{1}{15}\tau_c^{\frac{3}{2}} - \frac{1}{3}\tau_d^{\frac{3}{2}}\right), \tag{4.32}$$

and the triple intersections can be rewritten as

$$I_3 = 147\,D_a^3 + 1225\,D_b^3 + 225\,D_c^3 + 9\,D_d^3\,. \tag{4.33}$$

The Kähler cone is the subspace of the space of parameters t_i for which the condition $\int_C J > 0$ holds. In this case, the Kähler cone conditions are:

$$\begin{aligned}
t_1 - 2\,t_3 &> 0\,,\\
-2\,t_1 + t_2 + 3\,t_3 &> 0\,,\\
t_2 - 3\,t_4 &> 0\,,\\
2\left(t_3 - t_2\right) + t_4 &> 0\,.
\end{aligned} \tag{4.34}$$

Now that we have the volume (4.30) in explicit 'Swiss cheese' form, we can search for the large volume limit at which we would like to stabilize the CY. The idea is to find the right divisor η_l, such that when its volume τ_l grows, only τ_a will grow, and $\tau_b, \ldots, \tau_d$ will remain constant. In this case, η_l is clearly $\eta_3 = D_8$. Naïvely, we could declare our large volume limit to be

$$\text{Naively}:\quad \tau_3 \to \infty\,;\quad \tau_1, \tau_2, \tau_4 \quad \text{constant and small}\,. \tag{4.35}$$

By looking at the projective weights of the coordinates in table 4.1, we conclude that any divisor that is charged with respect to the third row will grow large, whereas any divisor that is not will remain constant in volume. Henceforth, we will refer to η_3 as a 'large direction', or 'large' divisor. However, care must be exercised in trying to shrink the so-called 'small' divisors. Although one would, by inspection of (4.30), conclude, that the directions τ_1, τ_2, and τ_4 can be shrunk to arbitrarily small size while keeping τ_3 arbitrarily large, a careful analysis of the Kähler cone conditions (4.34) reveals that this is not entirely possible. If we rewrite these conditions in terms of the divisor volumes as follows:

$$\begin{aligned} 7\sqrt{\tau_b} - 3\sqrt{\tau_c} &> 0\,, \\ 3\sqrt{\tau_d} &> 0\,, \\ \sqrt{\tau_a} - 5\sqrt{\tau_b} &> 0\,, \\ -\sqrt{\tau_a} + 5\sqrt{\tau_b} + 5\sqrt{\tau_c} - \sqrt{\tau_d} &> 0\,, \end{aligned} \qquad (4.36)$$

we see from the last condition that sending τ_a large forbids setting both τ_b and τ_c very small. At least one of these two volumes will have to be large. By carefully analyzing these conditions, we conclude that the only possible large volume limits are the following two:

$$\tau_1\,,\tau_4 \to 0\,, \quad \tau_2\,,\tau_3 \to \infty\,; \qquad (4.37)$$

$$\tau_2\,,\tau_4 \to 0\,, \quad \tau_1\,,\tau_3 \to \infty\,. \qquad (4.38)$$

As we will see in the next subsection, this phenomenon can be linked to the topology of the divisors.

Identifying smooth, 'small' cycles

We will now search for all smooth, potentially 'small', effective divisors in this model, on which we will subsequently wrap our MSSM branes and our E3-branes. We will require smoothness, in order to be able to reliably compute Hodge numbers and induced charges.

As explained in the previous section, any divisor that is not charged under the third $\mathbb{C}^*$ action shown in table 4.1 has at least the potential to be 'small'. In other words, such a divisor must be of the form $D = k\,\eta_1 + l\,\eta_2 + m\,\eta_4$. However, by inspecting the weight table, we see that such a divisor will always only have one monomial to represent it, namely

$$x_5^k\, x_6^l\, x_8^m\,. \qquad (4.39)$$

Hence, the only smooth (i.e. irreducible), small divisors are D_5, D_6, and D_8.

We would now like to compute the Hodge numbers of these three divisors. Given the fact that all three of them are rigid (i.e. have no deformations), Serre duality tells us that they have $h^{0,2} = 0$. In order to compute $h^{0,1}$, we will use the index formula for the holomorphic Euler characteristic (4.23). Plugging in the data for this CY, we find for $D = k\,\eta_1 + l\,\eta_2 + m\,\eta_4$

$$\begin{aligned} \chi(D, \mathcal{O}_D) =\ & -\frac{1}{3}k + \frac{4}{3}k^3 - \frac{1}{3}l + \frac{3}{2}k^2 l - \frac{5}{2}kl^2 + \frac{4}{3}l^3 \\ & + 4m - \frac{21}{2}k^2 m + \frac{45}{2}km^2 - 16\,m^3 - \frac{1}{2}n + \frac{1}{2}l^2 n - \frac{3}{2}ln^2 + \frac{3}{2}n^3\,. \end{aligned} \qquad (4.40)$$

Looking for a choice of parameters (k,l,m) such that $\chi(D,\mathcal{O}_D) = 1$ we find the solutions

$$(k,l,m) = \{(1,0,0)\,,(0,1,0)\,,(0,0,1)\,,(1,1,0)\,,(0,1,1)\,,(1,1,1)\}\,.$$

The last three divisors in the list are reducible, and hence not smooth. The first three are precisely the ones we identified before. This calculation shows that all three of them have

x_2	x_3	x_4	x_6
6	1	1	1
1	0	0	1

Table 4.2: Charges of the Hirzebruch surface $\mathbb{F}_5$.

$h^{0,1} = 0$, i.e. no Wilson lines. This means that these divisors are perfect for all our purposes: We want to avoid having extra neutral zero-modes on the instantons, we do not want to have D-branes with extra moduli to stabilize, and we want to be able to turn on NS-NS three-form flux without causing any Freed-Witten anomalies, all of which is avoided by having $h^{0,2} = b^1 = b^3 = 0$.

We can actually identify these divisors as *rational surfaces*. Rational surfaces are either Hirzebruch surfaces, $\mathbb{CP}^2$, or blow-ups of $\mathbb{CP}^2$ at up to eight points (i.e. del Pezzo surfaces). First of all, we notice that for all three D_5, D_6, D_8, the second plurigenus vanishes

$$p^2(D) \equiv \dim H^0(D, K_D^{\otimes 2}) = \dim H^0(D, N_D^{\otimes 2}) = 0, \qquad (4.41)$$

where N_D is the normal bundle of D. We see this by inspecting the table 4.1, and seeing that, for instance, a section of $N_{D_5}^{\otimes 2}$ would correspond to a monomial of class $2\eta_1$ that does not vanish on D_5. The only monomial in this class is x_5^2,[4] so there are no non-vanishing sections of this bundle. The same occurs for the other two divisors. The vanishing of the second plurigenus, plus the fact that $h^{0,1} = 0$, implies by the Castelnuovo-Enriques theorem (see section 4.4 of [67]) that these surfaces are rational. The Euler numbers of the three divisors are easily computed by means of the formula $\chi(D) = \int_X (D^3 + c_2(X) \wedge D)$ to be

$$\begin{aligned} \chi(\eta_1) &= 4, \quad \chi(\eta_2) = 4, \quad \chi(\eta_4) = 3, \\ \implies h^{1,1}(\eta_1) &= 2, \quad h^{1,1}(\eta_2) = 2, \quad h^{1,1}(\eta_4) = 1. \end{aligned} \qquad (4.42)$$

Let us take a closer look at D_5. By inspecting expression (4.26), we see that if $x_5 = 0$ then both x_1 and x_8 must be non-vanishing. Hence, we can gauge-fix both coordinates $x_1 = x_8 = 1$. This uses up two projective $\mathbb{C}^*$-actions. Let us choose the gauge fixing such that the first and the last rows of table 4.1 are eliminated. If we now write down the polynomial defining the CY, after setting $x_5 = 0$ and gauge-fixing, we have something of the form:

$$P(x_2, x_3, x_4, x_6, x_8) + x_7 = 0. \qquad (4.43)$$

Hence, x_7 is uniquely determined by the other coordinates, so we can eliminate it. After taking the appropriate linear combination of the charge rows, we are left with the toric description of the surface given in table 4.2. This is nothing other than the fifth Hirzebruch surface $\mathbb{F}_5$. It is not a del Pezzo surface, because its anticanonical bundle is not ample. In fact, we could have seen this more quickly by inspecting the intersection numbers (4.27). It is well known that an

---
[4]Checking this is not entirely trivial. One must also take the SR ideal in (4.26) into account. As certain monomials are not allowed to vanish on a surface, it is possible to build sections that are quotients of monomials.

ample line bundle on a surface has to have a positive intersection with any effective curve on the surface. In our case, this means that

$$\int_{D_5} (-K_{D_5}) \cdot C > 0 \quad \text{for any effective curve} \quad C \in D_5. \tag{4.44}$$

Taking the curve defined as $C: \{x_5 = 0\} \cap \{x_6 = 0\}$, we can compute

$$\begin{aligned} \int_{D_5} (-K_{D_5}) \cdot C &= \int_{D_5} (-\eta_1) \cdot C \\ &= \eta_1 \cdot (-\eta_1) \cdot \eta_2 = -3. \end{aligned} \tag{4.45}$$

Hence, we see that this surface can not be del Pezzo. Similarly, we see from the number $\eta_2^2 \eta_4 = +1$ that our second surface, D_6 must also be a non-del Pezzo Hirzebruch surface. This explains why we can not simultaneously shrink both of these surfaces arbitrarily as we naïvely would have expected. Our third surface, D_8, however, is simply a $\mathbb{CP}^2$, which is a Del Pezzo. It can be shrunk arbitrarily. Notice that D_8 is the only surface whose volume appears 'diagonally' in the volume function of the CY.

4.4.2 Scenarios in the first model

Step one: 'Local' models

We will study a setup with two stacks of D7-branes $D7_A$ and $D7_B$, each one on a different 'small' cycle, plus one E3-brane on another 'small' cycle. The reason for placing the MSSM on 'small' cycles, is to keep the gauge coupling constants large. We want two different D7 stacks in order to get chiral matter in four dimensions. We would like the MSSM gauge group to be unitary. There are two ways to accomplish this. One way would be to have D7/image-D7-brane configurations (as opposed to D7-branes on top of the O7-plane, or transversally involution invariant D7-branes). However, since the cycles we are dealing with are rigid, they are automatically left invariant by involutions of the type we consider in this chapter. The other way to get unitary gauge groups is to have transversally invariant, even ranked stacks, which will induce symplectic gauge groups, and then turn on a diagonal flux to break the latter to unitary groups.

The E3-brane on the other hand, must have an $O(1)$ gauge group. This is accomplished by having a single E3 placed on a transversally invariant cycle.

We have three possible cycles on which to place the E3-instanton. Having fixed that choice, the two MSSM branes will occupy the other two 'small' cycles. Let us begin by putting an E3 on η_1. The charge vector for this brane is

$$\Gamma_{E3} = \eta_1 + \tfrac{1}{2} \eta_1^2 + \tfrac{7}{6} \omega, \tag{4.46}$$

where the two-form, four-form and volume-form correspond to D3-, D1-, and D(-1)- charges, respectively. The half-integral four-form corresponds to the flux $F = \tfrac{1}{2} \eta_1$ that compensates for the Freed-Witten anomaly. The four-form can be geometrically interpreted as the Poincaré dual to the curve on which the induced D1 would be wrapped. However, if we integrate all possible basis elements of $H^2(X,\mathbb{Z})$ we obtain

$$\int_{E3} \tfrac{1}{2} \imath^*(\eta_1) \cdot \imath^*\{\eta_1,\eta_2,\eta_3,\eta_4\} = \int_X \tfrac{1}{2}\eta_1^2 \cdot \{\eta_1,\eta_2,\eta_3,\eta_4\}$$
$$= \{4, \tfrac{3}{2}, -\tfrac{21}{2}, 0\}. \qquad (4.47)$$

This curve is not a well-defined (integral) element of $H_2(X,\mathbb{Z})$. In other words, it fails the test we defined in section 4.2, which means that this half-integer flux can not be compensated by turning on more flux on the E3.

Due to the non-vanishing $U(1)$ world-volume field-strength $F = \tfrac{1}{2}\eta_1$, this E3-brane is not invariant under orientifolding. In order to fix this we must turn on an appropriate B-field[5]

$$B = F = \tfrac{1}{2}\eta_1, \qquad (4.48)$$

such that

$$\mathcal{F} = F - B = 0 = -\sigma^*(\mathcal{F}). \qquad (4.49)$$

Two comments are in order: first of all, notice, that since we now have $B = F$ on the E3-brane, the latter automatically satisfies the D-term constraint, i.e. has a vanishing FI parameter (A.6). Its central charge is aligned with that of the O7-plane. Since the B-field can not run continuously, this means that this instanton can not become non-BPS (unless, of course, supersymmetry is broken by the other branes present), and we do not have to worry about extra fermionic zero-modes appearing in different regions of the moduli space. As explained in [68, 69], this means that this instanton will contribute to the superpotential, as opposed to giving rise to the higher F-terms considered in [70, 71]. Secondly, having fixed the B-field at this value, it is now impossible for other instantons wrapped on the other two small cycles to contribute, as their respective Freed-Witten compensating fluxes differ from the one in this case.

Now, we move on to set up our MSSM D7-branes. We will do this in two stages. First, we will place two rank one D7-branes on the two remaining small divisors and tackle the problem of the unwanted E3-D7 strings.[6] In the next subsection, we will scan for involutions and try to embed the system into a consistent, global (tadpole canceling) model, and see whether we can still solve the problem of unwanted charged zero-modes and unwanted matter after we are forced to add tadpole canceling 'hidden' D7-branes.

We place two D7-branes, D7$_A$ and D7$_B$ on the remaining small divisors, η_4 and η_2, respectively. Both branes fail our test for the Freed-Witten flux, i.e. their FW fluxes can not be turned off. By inspecting (4.27), we see that η_1 and η_4 never intersect on the CY. Hence, there

---
[5]Note that, because we only consider involutions with $H^2_-(X,\mathbb{Z}) = 0$, the B-field is frozen. However, it is allowed to take on discrete values such that $B = -\sigma^*(B)$ mod $H^2(X,\mathbb{Z})$.

[6]As this problem is insensitive to the ranks of the stacks, we will set them to one for now, and adjust them later as needed. Everything we do now will carry over to the case of higher rank stacks. One just needs to take the tensor product of the line bundles we construct here with traceless vector bundles.

are no zero-modes charged under the $D7_A$. If we now compute the chiral intersection number between the E3 and the $D7_B$, as defined in equation (A.7), we find

$$\langle \Gamma_{E3}, \Gamma_B \rangle = 4 \,. \tag{4.50}$$

This will induce four unwanted charged zero-modes in four dimensions. Hence, we need to turn on extra flux on both branes to cancel this.

Let us define the added fluxes (on top of the half-integral fluxes) ΔF_B and ΔF_{E3} on the $D7_B$ and the E3 as follows:

$$\Delta F_B = \{b_1; b_2; b_3; b_4\}, \tag{4.51}$$
$$\Delta F_{E3} = \{e_1; e_2; e_3; e_4\}, \tag{4.52}$$

where the b_i's and e_i's are integer components with respect to the η_i basis, and we have suppressed the pullback symbol. Computing the charge vectors again we get the intersection number

$$\langle \Gamma_B, \Gamma_{E3} \rangle = 4 - 3 b_1 + 5 b_2 + 3 e_1 - 5 e_2 \,. \tag{4.53}$$

Setting this to zero yields the following seven-parameter solution:

$$\Delta F_B = \{3 + e_1 + 5 n; 1 + e_2 + 3 n; b_3; b_4\}, \tag{4.54}$$
$$\Delta F_{E3} = \{e_1; e_2; e_3; e_4\}, \tag{4.55}$$

where n is an arbitrary integer, as are the other parameters. In order to maintain the orientifold invariance of the E3-brane, the B-field must always be adjusted such that $B = \frac{1}{2} \eta_1 + \Delta F_{E3}$. The number of chiral bifundamental A-B-strings is then given by

$$\langle \Gamma_A, \Gamma_B \rangle = 3 (n + a_4 - b_4) - a_2 + e_2 \,. \tag{4.56}$$

For both the $D7_A$ and the $D7_B$ we get

$$\xi_A, \xi_B \sim (t_2 - 3 t_4) = \sqrt{\tau_4} \,. \tag{4.57}$$

Hence, these D-terms want to shrink the $D7_A$ to zero size, bringing us to the boundary of the Kähler cone. The formula used to compute these D-terms, however, is only valid at large radius. Once the cycle η_4 reaches stringy scale, world-sheet instanton corrections will dominate and drastically modify the central charge of the D-brane. Computing these corrections exactly would require solving the Picard-Fuchs equations for the mirror CY, which is beyond the scope of this work. On the other hand, it is more plausible that the cycles of both the $D7_A$ and $D7_B$ will get stabilized within the Kähler cone by string loop effects, as has been worked out in general in [50, 54].

Now, let us reshuffle the branes and place the E3, $D7_A$ and $D7_B$ on η_4, η_1 and η_2, respectively. We obtain the following solution:

$$\Delta F_A \quad \text{arbitrary}, \tag{4.58}$$
$$\Delta F_B = \{b_1; 1 + e_2 + 3 n; b_3; 1 + e_4 + n\}, \tag{4.59}$$
$$\Delta F_{E3} = \{e_1; e_2; e_3; e_4\}. \tag{4.60}$$

Scenario	E3	D7$_A$	D7$_B$
I	η_1	η_4	η_2
	arbitrary	arbitrary	$\{3+e_1+5n\,;\,1+e_2+3n\,;\,b_3\,;\,b_4\}$
II	η_4	η_1	η_2
	arbitrary	arbitrary	$\{b_1\,;\,1+e_2+3n\,;\,b_3\,;\,1+e_4+n\}$

Table 4.3: Two 'local' models.

This system has
$$\langle \Gamma_A, \Gamma_B \rangle = 3\left(-a_1 + b_1 - 5n - 3\right) + 5\left(a_2 - e_2\right), \tag{4.61}$$
bifundamental, chiral, A-B-strings.

Finally, we could now go on to reshuffle the branes again, but this would force us to put the MSSM branes on η_1 and η_4, which do not intersect at all. This would defeat the purpose of having a chiral MSSM setup. We summarize the results in table 4.3.

Step two: 'Global' models

By 'global' model, we will mean a model where an involution has been chosen, and all D7-charge has been cancelled. In the previous section we identified the divisors on which we want to wrap the instanton and two stacks of intersecting MSSM branes. However, such setups will typically not cancel the total D7-tadpole, and a third (set of) branes will have to be added. It is phenomenologically desirable that these new branes do not intersect the MSSM branes, nor the E3-brane. In this section, we will find out to what extent it is possible to solve this problem.

Let us begin with scenario I in table 4.3, where the two stacks have ranks N_A and N_B. We will pick an involution and explain the procedure by working out the example. Define the involution as
$$x_1 \longrightarrow -x_1. \tag{4.62}$$
The resulting O7-plane has D7-charge
$$-8 \times [O7] = -(120\,\eta_1 + 72\,\eta_2 + 56\,\eta_3 + 24\,\eta_4). \tag{4.63}$$
Taking into account the D7$_A$ and D7$_B$ with their arbitrary ranks, N_A, N_B and their image-branes, means that we have to make up for
$$120\,\eta_1 + (72 - N_B)\,\eta_2 + 56\,\eta_3 + (24 - 2\,N_A)\,\eta_4 \tag{4.64}$$
worth of D7-charge. This charge can be distributed in many ways: we can distribute it among several branes or use just one brane; we can use brane/image-brane pairs, or Whitney-type branes (see section A.1.2). Let us first address the question as to whether one should distribute the charge among several branes, or just a single brane. Picking a single brane with the full charge in (4.64) has several advantages over partitioning the charge among more branes. First

of all, the hidden brane has to have zero intersection product with the E3, the D7$_A$ and the D7$_B$. If we were to partition the hidden brane into several branes,

$$\Gamma_H = \sum_i \Gamma_{H_i}, \tag{4.65}$$

with each Γ_{H_i} satisfying the zero intersection property, then the sum Γ_H would also satisfy it. It is a necessary condition that the total charge satisfy it, in order to solve the problem for the constituents. It is therefore much simpler to only have to solve this problem once for one brane. The second advantage lies in the fact that a single high charge D7-brane will typically generate a much larger curvature induced D3-charge than several low charge branes. Schematically, in a one-modulus CY, the Euler number of a degree N divisor grows like $\sim N^3$, whereas N degree one branes will simply induce a total charge N. We will therefore work with a single hidden brane.

The second question concerns the type of brane we should use. We claim that it is more advantageous to use a Whitney-type hidden brane. Whitney-type branes are invariant under the involution by construction. This means that the E3 is automatically orthogonal to it. This also means that imposing that the hidden brane has to be orthogonal to the D7$_A$ and D7$_B$ automatically makes it orthogonal to their respective images. Finally, the invariance means that the hidden brane automatically has a trivial D-term. The D-term for a non-invariant hidden brane, which is always wrapped on a large cycle, would typically ruin the large volume limit. This can be understood as follows. The charge vector of the hidden brane is orthogonal to those of the E3, the two MSSM branes, and their respective images. Combining these equations, and using the diagonal basis from (4.31) one can show that

$$D_H \cdot D_{E3,D7_A,D7_B} \cdot (F_H - B) = 0 \implies D_H \cdot D_{b,c,d} \cdot (F_H - B) = 0. \tag{4.66}$$

Hence, the FI-term for the hidden brane will necessarily be proportional to $\tau_a \sim t_3$, i.e. the volume of the large cycle. This would force the CY to be small. A Whitney-type brane automatically circumvents this problem.

Therefore, we will search for a single Whitney-type D7-brane of charge $[D_W] = 2\,[D_P]$ given by (4.64). The easiest way to construct its charge vector is by using the K-theoretic picture, as described in [72] and summarized in appendix A.1.3. As explained there, the choice of shift flux does not enter the intersection numbers between the Whitney brane W and the other branes present. All to do is to solve the equations:

$$\langle \Gamma_H, \Gamma_A \rangle = \langle \Gamma_H, \Gamma_B \rangle = 0. \tag{4.67}$$

The solution is simply $N_B = 3\,N_A$. This means that this scenario can generate models with gauge groups of the form $U(3\,N) \times U(N)$. Let us now compute the tadpole that our hidden brane generates. For this computation, we will have to assign a value to the shift vector. For simplicity, let us choose $N_B = 3, N_A = 1$. The constraints on the shift flux S from (A.24) become

$$\{52 - e_1, \tfrac{57}{2} - e_2, \tfrac{49}{2} - e_3, \tfrac{19}{2} - e_4\} \geq S \geq \{7 - e_1, \tfrac{9}{2} - e_2, \tfrac{7}{2} - e_3, \tfrac{3}{2} - e_4\}. \tag{4.68}$$

Notice that we can not saturate these constraints. This might indicate the presence of a flux on the brane that can not be switched off due to some anomaly. Let us choose S to be 'minimal':

$$S = \{7 - e_1, 5 - e_2, 4 - e_3, 2 - e_4\}. \tag{4.69}$$

Now we may compute the 'physical' (gauge invariant) D3 tadpole by taking the six-form component as follows:

$$\left(\Gamma_W\, e^{-B}\right)_{6-\text{form}} = \frac{7763}{4} \approx 1940. \tag{4.70}$$

Let us now repeat this calculation for the second scenario. Starting with arbitrary ranks N_A and N_B again, we solve the equations

$$\langle \Gamma_W, \Gamma_A \rangle = 0, \qquad \langle \Gamma_W, \Gamma_B \rangle = 0. \tag{4.71}$$

The second equation is proportional to $3\,N_A - 5\,N_B$. Setting it to zero and eliminating N_A in the first equation yields a term proportional to N_B. This means, we can not choose non-zero ranks and turn off the intersections with the hidden brane. Therefore, the second scenario has a visible 'hidden' sector.

4.4.3 Moduli stabilization analysis

From equation (4.13) we see that as long as the magnetized D7-branes are small the potential is of order $1/\mathcal{V}^2$, so the D-term part will dominate over the F-term contribution in the LVS. The curvature along its non-flat directions is much larger than the one of the F-term potential. In the limit of exponentially large volume (the divisor of the D7-brane has to remain small) this generates an exponentially strong force in comparison to the F-term forces. Hence, in the following, we will use $V_D = 0$ as a constraint on our configuration and just look at the F-term potential.

To obtain the concrete form of the F-term potential (4.12) for our scenarios we have to calculate the self-intersection volume for the instanton. In the first case it is given by

$$\text{Vol}\left(D_{E3} \cap D_{E3}\right) = 8t_1 + 3t_2 - 21t_3 = -\frac{\sqrt{2}}{5}\left(7\sqrt{\tau_b} + 3\sqrt{\tau_c}\right), \tag{4.72}$$

and in the second one we obtain

$$\text{Vol}\left(D_{E3} \cap D_{E3}\right) = -3t_2 + 9t_4 = -3\sqrt{2}\sqrt{\tau_d}. \tag{4.73}$$

Knowing these, we can write the potentials as a function depending on τ_a, τ_c, τ_d and τ_{E3}. So for the first scenario we find

$$\begin{aligned}V_F &= \frac{1}{\hat{\mathcal{V}}^2}\Big(\frac{\sqrt{2}}{5} 4\pi^2 \left(7\sqrt{5\tau_{E3} + \tau_c} + 3\sqrt{\tau_c}\right) \hat{\mathcal{V}} |A_{E3}|^2\, e^{-4\pi\hat{\tau}_{E3}} \\ &\quad - 4\pi\hat{\tau}_{E3} e^{-2\pi\hat{\tau}_{E3}} |A_{E3} W_0| + \frac{3}{4}\frac{\hat{\xi}}{\hat{\mathcal{V}}}|W_0|^2\Big),\end{aligned} \tag{4.74}$$

where $\hat{\mathcal{V}}$ is also a function of the divisor volumes above. Now we search for a minimum of the potential

$$dV_F = 0 \quad \Longrightarrow \quad \frac{\partial V_F}{\partial \tau_d} = \frac{\partial V_F}{\partial \hat{\mathcal{V}}} \frac{\partial \hat{\mathcal{V}}}{\partial \tau_d} = 0. \tag{4.75}$$

Hence, we can also solve $\frac{\partial V_F}{\partial \hat{\mathcal{V}}} = 0$.

$$\Longrightarrow \quad \mathcal{V} = \frac{5 g_s W_0 \tau_{E3} e^{\frac{2\pi \tau_{E3}}{g_s}}}{\sqrt{2} A_{E3} \pi (3\tau_c + 7\sqrt{5\tau_{E3} + \tau_c})}$$

$$\pm \frac{\sqrt{5} g_s W_0 e^{\frac{2\pi \tau_{E3}}{g_s}} \sqrt{80 \tau_{E3}^2 - 9\sqrt{2}\xi(3\sqrt{\tau_c} + 7\sqrt{5\tau_{E3} + \tau_c})}}{4\sqrt{2} A_{E3} \pi (3\tau_c + 7\sqrt{5\tau_{E3} + \tau_c})}. \tag{4.76}$$

What is important here is that, although the potential looks like the one in [4], there is a subtle difference to our case. If we demand a large volume while at the same time fulfilling the Kähler cone constraints, the term in the square root becomes negative. Hence we can not realize the desired LVS in scenario I.[7]

Let us now look at the second scenario. Here we obtain the same form for the potential as [3]:

$$V_F = \frac{1}{\hat{\mathcal{V}}^2} \left(12\pi^2 \sqrt{2\tau_{E3}} \hat{\mathcal{V}} |A_{E3}|^2 e^{-4\pi\hat{\tau}_{E3}} - 4\pi \hat{\tau}_{E3} e^{-2\pi\hat{\tau}_{E3}} |A_{E3} W_0| + \frac{3}{4} \frac{\hat{\xi}}{\hat{\mathcal{V}}} |W_0|^2 \right), \tag{4.77}$$

so the only thing that could prevent us from having large volume stabilization are the Kähler cone (KC) conditions. This means that in the second scenario, although we could not solve the intersection problem, we can at least realize the LVS. Looking at the allowed large volume limits in (4.38), we can pick the case where τ_2 and τ_4 are small. In this case, we have to drop the D7$_A$-brane, since it would generate an FI-term of the form $\sqrt{5\tau_1 + 3\tau_2 + \tau_4}$, which would be fatal to this LVS. Choosing $|A_{E3}| = 1$, $|W_0| = 5$ and $g_s = \frac{1}{10}$ we find the following values for the CY and instanton volume at the location of the minimum

$$\tau_{E3} = 2.15,$$
$$\mathcal{V} = 1.46514 \cdot 10^{57}. \tag{4.78}$$

Note, however, that the FI-term generated by the remaining D7$_B$-brane, which is of the form $\sqrt{3\tau_2 + \tau_4}$, would naïvely force the instanton cycle to zero size, thereby destroying this LVS. It is, however, possible that the string loop corrections considered in [54] might counter this effect and keep the instanton size finite. But this is beyond the scope of our work.

The fact that the volume of η_4 appears diagonally in the CY volume is enough to get the right form of the F-term potential. One can also show that this surface actually resolves a point-like singularity. This is another affirmation of the theorem given in [50].

---

[7]This confirms the theorem of [50] that one obtains a minimum at exponentially large volume only if the instanton is wrapped around a local blow-up mode resolving a point-like singularity.

	x_1	x_2	x_3	x_4	x_5	x_6	x_7	x_8	p
	2	1	6	1	2	0	0	0	12
	2	1	5	0	2	0	0	2	12
	2	0	5	1	2	0	2	0	12
	1	0	3	0	1	1	0	0	6

Table 4.4: Projective weights for the R1 resolution of $\mathbb{P}^4_{2,1,6,1,2}(12)/\mathbb{Z}_2 : 1\,0\,0\,0\,1$.

4.5 Second model

The results for all of our scenarios throughout this chapter are concisely summarized in table 4.6.

4.5.1 R1 resolution of $\mathbb{P}^4_{2,1,6,1,2}(12)/\mathbb{Z}_2 : 1\,0\,0\,0\,1$ geometry

Our next model is the first of two resolutions of the orbifolded weighted projective space $\mathbb{P}^4_{2,1,6,1,2}(12)/\mathbb{Z}_2 : 1\,0\,0\,0\,1$. Here, the integers $(1,0,0,0,1)$ denote the charges of the coordinates under the $\mathbb{Z}_2$-action. The projective weights for this model are listed in table 4.4.

The Stanley-Reisner ideal of the ambient space reads

$$I_{SR} = \{x_2x_3,\ x_2x_4,\ x_3x_4,\ x_3x_6,\ x_4x_7,\ x_2x_8,\ x_1x_5x_6x_7,\ x_1x_5x_6x_8,\ x_1x_5x_7x_8\}. \tag{4.79}$$

The triple intersection numbers in the basis $\eta_1 = D_2$, $\eta_2 = D_4$, $\eta_3 = D_6$, $\eta_4 = D_8$ are encoded in

$$\begin{aligned} I_3 &= -78\eta_4^3 - 6\eta_3\eta_4^2 - 6\eta_3^2\eta_4 + 2\eta_3^3 + 36\eta_2\eta_4^2 + 6\eta_2\eta_3\eta_4 + \eta_2\eta_3^2 \\ &\quad -18\eta_2^2\eta_4 - 3\eta_2^2\eta_3 + 9\eta_2^3 + \eta_1\eta_3^2 - 3\eta_1^2\eta_3 + 9\eta_1^3\,. \end{aligned} \tag{4.80}$$

The Kähler form in the basis $\{\eta_1, \eta_2, \eta_3, \eta_4\}$ is given by

$$J = t_1\,\eta_1 + t_2\eta_2 + t_3\,\eta_3 + t_4\,\eta_4\,. \tag{4.81}$$

The volumes of the corresponding divisors are

$$\begin{aligned} \tau_1 &= \frac{1}{2}(-3t_1 + t_3)^2\,, \\ \tau_2 &= \frac{1}{2}(-3t_2 + t_3 + 6t_4)^2\,, \\ \tau_3 &= \frac{1}{2}\left(-3t_1^2 - 3t_2^2 + 2t_1t_3 + 2t_2t_3 + 2t_3^2 + 12t_2t_4 - 12t_3t_4 - 6t_4^2\right)\,, \\ \tau_4 &= \frac{1}{2}\left(-18t_2^2 + 12t_2t_3 - 6t_3^2 + 72t_2t_4 - 12t_3t_4 - 78t_4^2\right)\,. \end{aligned} \tag{4.82}$$

The volume of the CY manifold is given by

$$\begin{aligned} \mathcal{V} &= \frac{1}{18}\left[9t_4^3 - (-2t_3 + 3t_4)^3 - (-3t_1 + t_3)^3 - (-3t_2 + t_3 + 6t_4)^3\right] \\ &= \frac{\sqrt{2}}{9}\left[\frac{3}{2\sqrt{6}}(\tau_1 + 5\tau_2 + 3\tau_3 + 2\tau_4)^{\frac{3}{2}} - \frac{1}{2\sqrt{2}}(\tau_1 + \tau_2 + 3\tau_3)^{\frac{3}{2}} - \tau_1^{\frac{3}{2}} - \tau_2^{\frac{3}{2}}\right]\,. \end{aligned} \tag{4.83}$$

Scenario	E3	D7$_A$	D7$_B$	
I	η_1	η_2	η_3	
	arbitrary	arbitrary	$\{b_1 \,;\, b_2 \,;\, -2 + 3\,(b_1 - e_1) + e_3 \,;\, b_4\}$	
II	η_2	η_1	η_3	
	arbitrary	arbitrary	$\{b_1 \,;\, b_2 \,;\, -2 + e_3 + 3\,(-2\,b_4 - e_2 + 2\,e_4 + b_2) \,;\, b_4\}$	

Table 4.5: Two 'local' models.

It has the expected Swiss cheese form. From this volume formula we deduce the diagonal basis to be

$$\begin{aligned} D_a &= \eta_1 + 5\eta_2 + 3\eta_3 + 2\eta_4\,, \\ D_b &= \eta_1 + \eta_2 + 3\eta_3\,, \\ D_c &= \eta_1\,, \\ D_d &= \eta_2\,, \end{aligned} \qquad (4.84)$$

and the triple intersections can be rewritten as

$$I_3 = 24\,D_a^3 + 72\,D_b^3 + 9\,D_c^3 + 9\,D_d^3\,. \qquad (4.85)$$

The Kähler cone conditions are:

$$\begin{aligned} t_3 - t_4 &> 0\,, \\ t_2 - t_3 - t_4 &> 0\,, \\ t_1 - t_3 + t_4 &> 0\,, \\ -3t_2 + t_3 + 6t_4 &> 0\,, \\ -3t_1 + t_3 &> 0\,. \end{aligned} \qquad (4.86)$$

For this model, the 'small', rigid cycles with holomorphic Euler characteristic are

$$\{D_2,\, D_4,\, D_6\} = \{\eta_1,\, \eta_2,\, \eta_3\} \quad \text{with} \quad h^{1,1} = \{1, 1, 8\}\,. \qquad (4.87)$$

The first two surfaces are necessarily $\mathbb{CP}^2$'s. The third one, however, can not be a del Pezzo due to the intersection number $\eta_3^2\,\eta_2 = +1$, which implies that the anti-canonical bundle is not ample. This means that this surface must be a blow-up of $\mathbb{CP}^2$ at seven points that are not in generic position. In appendix A.2, we will explicitly work out one such 'pathological' surface that also fails to be a del Pezzo.

4.5.2 Scenarios in the second model

The divisors η_1 and η_2 do not intersect, therefore, we again only have two possible scenarios, which we summarize in table 4.5.

Let us move on to the global analysis. We pick, for convenience, the involution $x_3 \to -x_3$. Solving the equations

$$\langle \Gamma_W, \Gamma_A \rangle = \langle \Gamma_W, \Gamma_B \rangle = 0, \tag{4.88}$$

we find the following solutions.

1. **Scenario I**: the constraints we get from setting the chiral intersections with the hidden brane to zero are the following:

$$N_A = 3 N_B, \tag{4.89}$$

$$b_4 = -1 + 2(b_1 - e_2) + e_4. \tag{4.90}$$

We again have a setup that requires further constraints on the 'local' model. This time, however, these constraints are particularly simple to solve. To get an idea of how much D3-tadpole this Whitney-type brane can induce, let us compute it for the 'minimal' choice of the shift vector S in formula (A.24):

$$Q_{W,D3} = 372 - \frac{3}{2} N_A - 21 N_A^3. \tag{4.91}$$

This function bears a striking similarity with the results found in the previous model in appendix A.2. Finally, let us compute the FI-terms for both MSSM branes in light of these constraints:

$$\xi_A, \xi_B \propto \sqrt{\tau_2}. \tag{4.92}$$

The self-intersection volume for the instanton in this scenario is given by

$$\mathrm{Vol}\,(D_{E3} \cap D_{E3}) = 9t_1 - 3t_3 = -3\sqrt{2}\sqrt{\tau_c} = -3\sqrt{2}\sqrt{\tau_1}. \tag{4.93}$$

Writing the Kähler cone in the diagonal basis yields

$$\begin{aligned}
\sqrt{\tau_a} - 3\sqrt{\tau_b} &> 0, \\
2\sqrt{\tau_b} - \sqrt{\tau_d} &> 0, \\
2\sqrt{\tau_b} - \sqrt{\tau_c} &> 0, \\
\sqrt{\tau_d} &> 0, \\
\sqrt{\tau_c} &> 0.
\end{aligned} \tag{4.94}$$

Note that these conditions imply that we are free to make all three cycles η_1, η_2, η_3 small and still have a large volume limit where τ_4 is kept large. We observe that in this scenario the D-term forces us to the boundary of the Kähler cone. Relaxing the Kähler cone relations a bit and allowing non-strict inequalities, we will use $\tau_2 = 0$ as a constraint in the following. The F-term potential takes the form of (4.77) with the minimum

$$\tau_{E3} = \tau_1 = 1.25, \quad \mathcal{V} = 2.5945 \cdot 10^{32}. \tag{4.95}$$

Scenario	1 (I)	1 (II)	2 (I)	2 (II)	3 (I)	3 (II)	4 (I)	4 (II)
Global	✓	×	✓	✓	✓	✓	×	×
St. mod.	×	×	3	3	×	×	3	3

Table 4.6: Summary of results. The labels represent the model numbers and scenario numbers. For each scenario we indicate with a ✓ or a ×, whether the 'global problem' of suppressing undesirable intersections while canceling the D7 tadpole was solved. We also indicate how many, if any, Kähler moduli were successfully stabilized in each scenario.

2. **Scenario II**: the constraints for this scenario are the following:

$$N_A = 3 N_B, \qquad (4.96)$$
$$b_4 = 2n + 1 + e_4, \quad \text{for} \quad n \in \mathbb{Z}, \qquad (4.97)$$
$$b_2 = 5n + 3 + e_2. \qquad (4.98)$$

Let us also compute the D3 tadpole for this hidden brane with the 'minimal' choice of S:

$$Q_{W,D3} = 372 + \frac{3}{2} N_A - 75 N_A^3. \qquad (4.99)$$

The function is identical to that of the first scenario. In this case, both branes give again similar FI-terms:

$$\xi_A, \xi_B \propto \sqrt{\tau_1}. \qquad (4.100)$$

In this scenario, the self-intersection volume is

$$\text{Vol}(D_{E3} \cap D_{E3}) = 9t_2 - 3t_3 - 18t_4 = -3\sqrt{2}\sqrt{\tau_d} = -3\sqrt{2}\sqrt{\tau_2}. \qquad (4.101)$$

Here, we are again forced to the boundary of the Kähler cone. Again, relaxing the strict inequalities, we impose $\tau_1 = 0$. The F-term potential takes the form of (4.77) with the minimum:

$$\tau_{E3} = \tau_2 = 1.25, \quad \mathcal{V} = 2.5945 \cdot 10^{32}. \qquad (4.102)$$

Thus in the LVS of this model we were able to stabilize in both scenarios three out of the four Kähler moduli and again we expect that one can stabilize the last modulus via string loop corrections [50]. Note that both scenarios here yield the same potentials and the same values for the volumes. This is possibly due to the fact that both divisors η_1 and η_2 have the same topology: both are $\mathbb{CP}^2$'s. In fact, we notice in table (4.4) and in (4.79) that the CY threefold is symmetric under the simultaneous exchanges

$$x_2 \longleftrightarrow x_4, \qquad x_7 \longleftrightarrow x_8. \qquad (4.103)$$

4.6 Summary and outlook

In this chapter, we have searched for realizations of the Large Volume Scenario that are compatible with the presence of MSSM D7-branes with chiral matter, in the sense explained in [4].

We found that it is necessary to have at least three cycles that contribute negatively to the CY volume: two on which we placed two D7-stacks, and one for the E3-brane.

For this purpose we searched the list of 1197 toric CY hypersurfaces with $h^{1,1} = 4$. For simplicity we started with the 11 simplicial polytopes, which correspond to weighted projective spaces or quotients thereof. An extension of the package PALP [1,2] has been used to triangulate the 8 polytopes for which all divisors on the CY are toric and to compute the Mori cone and the intersection rings; see chapter 2. We thus found four inequivalent CYs of large-volume type.

Properly taking into account the fact that the Freed-Witten anomaly forces most of the branes (both D7 and E3) to be magnetized, we found that requiring vanishing chiral intersections between the E3 and the D7's, and between the 'hidden' D7 and the rest of the setup is stringent enough to rule out some of these models entirely. Throughout this chapter, we used the representation of D7-branes in terms of D9-anti-D9 condensates, which simplifies calculations of both induced charges and chiral intersections greatly. We did not specifically count vector-like pairs of chiral modes, but this can easily be done by literally counting sections of the appropriate bundles as opposed to using index theory. These issues were also carefully considered in [73].

For each model, we analyzed the topologies of the rigid, complex surfaces. We found that not all surfaces that are 'small', in the sense that they contribute negatively to the CY volume, are also del Pezzo. We found, as expected, that only the surfaces that *are* del Pezzo can be shrunk arbitrarily without spoiling the desired LVS. This means, that some Kähler moduli can not be stabilized by instanton effects or by D7-brane induced D-term constraints. Further analysis is needed to determine, whether string loop corrections [54] can lift those flat directions.

Our approach can be seen as complementary to one of the approaches presented in [73]. We are searching for CY's with del Pezzos in them by searching for polytopes with the right properties. One of the several approaches of [73], which is based on the techniques developed in [74], on the other hand, was to start with a simple CY, i.e. the quintic, and subject it to del Pezzo transitions, thereby designing the desired divisor structure.

So far, our search has only yielded $\mathbb{CP}^2$ surfaces as true del Pezzo surfaces. This chapter should be considered as step one in the search for candidate 'Swiss cheese' CY's. We expect that taking into account the general rules laid out in [50], combined with the techniques presented here, will lead to viable models. Indeed, very encouraging and promising results have been recently achieved by the authors of [75]; they have proposed three MSSM-like chiral models in which all moduli are stabilized taking proper account of tadpole and Freed-Witten anomaly cancellation among other things.

Our results are summarized in table 4.6. For each scenario of each of the four CY's, we state whether the 'global' problem of setting up E3- and D7-branes such that only wanted intersections are present, and such that the D7-tadpole is canceled, is solved. For each scenario we also give the number of Kähler moduli that were successfully stabilized. If no LVS was possible, we put a cross in the slot.

In conclusion, we have shown that even with the more stringent conditions imposed by the

Freed-Witten anomaly it is still possible to combine the LVS with setups of chirally intersecting D7-branes. The constraints help rule out some models, but still allow for flexibility. We have also demonstrated that the use of Whitney-type branes is preferable to the use of the more familiar stacks of brane/image-brane pairs, whenever possible, because the former do not produce unwanted chiral intersections, and they induce a lot more of the desired D3-charge.

Chapter 5

Toric constructions of global F-theory GUTs

Starting with [76–78], F-theory has been recognized as a setup to elegantly construct Grand Unified Theories (GUTs) in string theory. The GUT model is localized on a 7-brane S inside a complex three-dimensional manifold B which is the base of a compact elliptically fibered Calabi-Yau fourfold X_4. Requiring a decoupling limit between gauge and gravity degrees of freedom makes it possible to discuss many questions in a gauge theory that captures the physics in the vicinity of the GUT brane S. These local F-theory GUTs have a rich yet simple structure which allows to analyze many phenomenological questions in remarkable detail. See for instance [79] for a review. Due to the localization of gauge degrees of freedom on the 7-brane, in contrast to GUT theories coming from the heterotic string, F-theory provides a framework for a bottom-up approach to constructing realistic models from string theory. There, the first priority is to work out the phenomenological details of a model without worrying about the full string compactification. While the success of this approach speaks for itself, it is necessary to connect the bottom-up results with top-down constructions where the paradigm is to find a consistent string compactification which can ideally accommodate all the features of the local models. Finding and understanding global F-theory models has recently received increased attention.

There are several reasons to consider a full F-theory compactification on an elliptically fibered Calabi-Yau fourfold. The obvious reason is of course that there are issues which cannot be addressed in local models, most notably monodromies, fluxes and anomaly cancellation. These questions have been addressed recently in [80–94]. Another motivation, which will be the central concern of our work, is to explicitly construct compact Calabi-Yau fourfolds and to check whether they are suitable for F-theory model building. This is necessary in order to show whether the realistic models coming from a local construction have an embedding in a string compactification. Furthermore, we wanted to build a database of examples which contains the data necessary for GUT model building.

The main goal of this chapter is to give a systematic construction of a particular class of fourfold geometries and to analyze them in view of F-theory model building. Since a full classification of Calabi-Yau fourfolds, including the subset of elliptically fibered ones, is not available we aim to provide a set of examples within a well-defined framework. Toric geometry is a valuable and versatile mathematical tool for constructing Calabi-Yau manifolds. A prescription to use toric geometry to construct global F-theory GUTs has been given in [81] and further elaborated on in [83]. See also [13] for a recent review article and [85] for a closely related construction. The general idea is the following: first, find a base manifold B which is a blow-up of a Fano hypersurface in $\mathbb{P}^4$. In a second step, obtain a Calabi-Yau fourfold by constructing an elliptic fibration over the base B. This Calabi-Yau is then a complete intersection of two hypersurfaces in a six-dimensional toric ambient space. In [95] a class of models has been worked out where the base manifold B is a Fano hypersurface in $\mathbb{P}^4$ with up to three point or curve blow-ups. This extended the set of examples given in [81,83] but the geometries were still in a very restricted class. For instance, no examples in a general weighted projective space had been considered. In this chapter, we will systematically construct this more general type of models. The present extension allows us, for example, to set up global F-theory GUTs on dP_8's that have not been found in the previous investigations.

In order to find more general fourfold geometries we look at the construction of [81] from a slightly different point of view. Instead of considering blow-ups of Fano threefolds, we pick a subset of 1088 of the 473 800 776 reflexive polyhedra in four dimensions [30]. These polyhedra describe toric ambient spaces for Calabi-Yau threefolds. In contrast to looking at the Calabi-Yau case, we consider hypersurfaces in these toric ambient spaces that have homogeneous equations with multidegree smaller than in the Calabi-Yau case. This will define the base manifold B. The elliptically fibered Calabi-Yau fourfolds can be constructed from the base data using standard tools in toric geometry. In our computer-based search for models we have made extensive use of the software package PALP [1, 2]. In total we have found 569 674 base geometries.

Having constructed the geometries is only the first step of the program. Step two is to filter out those models which are usable in F-theory model building. Our goal was to formulate some elementary and general constraints that can be phrased in the toric language. These constraints can be divided up into conditions on the base geometry and conditions on the fourfold. While the former are specific to F-theory model building, the latter are of a more technical nature. As for the base manifolds, the first constraint is regularity. Hypersurfaces that are not Calabi-Yau may inherit the singularities of the toric ambient space. One sufficient criterion for regular hypersurfaces, which can be examined using toric methods, is base point freedom: given an empty base locus, any point-like singularity of the ambient space can be avoided by a generic choice of the hypersurface equation. We can impose further constraints on the toric divisors of the base B. Since we would like to construct F-theory models on these divisors, del Pezzo surfaces are particularly interesting. In local F-theory GUTs the del Pezzo condition guarantees a decoupling limit. Furthermore, certain vanishing theorems avoid exotic matter in $SU(5)$ GUTs [78]. For global models decoupling limits are more subtle and yield further constraints on the base geometries. The conditions on the complete intersection Calabi-Yau fourfold are more elementary. In order to be able to use the tools of toric geometry, we restrict to those examples where the Calabi-Yau data is encoded in a reflexive lattice polytope and where the information about hypersurface equations is given by a nef-partition. In our construction it is not automatic that the nef-partition is compatible with the elliptic fibration over the base B. Another issue is the reflexivity of the polytope that encodes the toric data the fourfold. A majority of the fourfolds we have constructed is not described in terms of reflexive polytopes. Reflexivity is important for mirror symmetry but since this is not required in our setup Calabi-Yau fourfolds coming from non-reflexive polytopes may be interesting to look at. However, we lack several mathematical and computational tools to deal with them, which is why we have to exclude them in our discussion. Finally, there is unfortunately also a computational constraint: since the lattice polytopes for Calabi-Yau fourfolds can be quite large, a fair amount of models cannot be analyzed due to numerical overflows and long calculation times.

Having reduced the number of interesting models by the constraints above we can explicitly construct F-theory GUTs using the prescription of [81]. We will focus on $SU(5)$ and $SO(10)$ GUTs and analyze some basic properties such as genera of matter curves and the number of Yukawa couplings. We will also construct $U(1)$-restricted models as introduced in [87].

This chapter is structured as follows. In section 5.2, we analyze the geometries we have constructed. Furthermore, we discuss some examples and comment on the discrepancy of Euler numbers between the toric calculation and a formula given in [81]. A match between the Euler numbers obtained from toric geometry and those obtained from the formula of [81] indicates that a local description of the gauge fluxes in terms of the spectral cover construction is plausible. Section 5.3 is reserved for conclusions and outlook.

5.1 Construction of global models

In this section we explain how to construct global F-theory models. In section 5.1.1, we recall the basic structure of global F-theory GUTs. In section 5.1.2, we describe how to systematically construct the base manifolds B as hypersurfaces in toric ambient spaces. Furthermore, we discuss the properties of GUT divisors in B. Finally, section 5.1.3 is devoted to the elliptically fibered Calabi-Yau fourfolds.

5.1.1 Setup

The class of global F-theory models, we aim to construct, have been first introduced in [81]. The Calabi-Yau fourfolds are complete intersections of two hypersurfaces in a six-dimensional toric ambient space. Schematically, these equations have the following form:

$$P_B(y_i, w) = 0\,, \qquad P_W(x, y, z, y_i, w) = 0\,. \tag{5.1}$$

The first equation only depends on the coordinates (y_i, w) of the base of the fibration. Here we have singled out one coordinate w, indicating that the divisor S, defined by $w = 0$, is wrapped by the 7-brane which supports the GUT theory. The second equation in (5.1) defines a Weierstrass model, where (x, y, z) are the coordinates of the $\mathbb{P}_{231}$ fiber. For this type of elliptic fibrations P_W has a Tate form which is globally defined:

$$P_W = x^3 - y^2 + xyza_1 + x^2z^2a_2 + yz^3a_3 + xz^4a_4 + z^6a_6, \qquad (5.2)$$

where the $a_n(y_i, w)$ are sections of K_B^{-n} and x and y are section of K_B^{-2} and K_B^{-3}, respectively. Constructing a Tate model is only the first step on the way to a F-theory GUT model. In order for the divisor $w = 0$ to support the desired gauge group the sections $a_n(y_i, w)$ have to have a particular structure. Via Kodaira's classification [96] and Tate's algorithm [97] the base-coordinate dependent coefficients a_i in the Tate equation must factorize in a particular way with respect to w. In the following we will focus on the gauge groups $SU(5)$ and $SO(10)$. For $SU(5)$ we must have:

$$a_1 = b_5 w^0, \quad a_2 = b_4 w^1, \quad a_3 = b_3 w^2, \quad a_4 = b_2 w^3, \quad a_6 = b_0 w^5. \qquad (5.3)$$

An $SO(10)$ model is specified as follows:

$$a_1 = b_5 w^1, \quad a_2 = b_4 w^1, \quad a_3 = b_3 w^2, \quad a_4 = b_2 w^3, \quad a_6 = b_0 w^5. \qquad (5.4)$$

The b_is are sections of some appropriate line bundle over B that have at least one term independent of w.

Matter arises along curves inside the base manifold at loci where a rank 1 enhancement of the GUT group takes place. In $SU(5)$ F-theory GUTs the matter curves are at the following loci inside S:

$$\begin{aligned} b_3^2 b_4 - b_2 b_3 b_5 + b_0 b_5^3 = 0 & \quad \text{5 matter} \quad SU(6) \text{ enhancement,} \\ b_5 = 0 & \quad \text{10 matter} \quad SO(10) \text{ enhancement.} \end{aligned} \qquad (5.5)$$

The matter curves for the $SO(10)$ models are at:

$$\begin{aligned} b_3 = 0 & \quad \text{10 matter} \quad SO(12) \text{ enhancement,} \\ b_4 = 0 & \quad \text{16 matter} \quad E_6 \text{ enhancement.} \end{aligned} \qquad (5.6)$$

Yukawa couplings arise at points inside B where the GUT singularity has a rank 2 enhancement. In $SU(5)$ models the Yukawa points sit at:

$$\begin{aligned} b_4 = 0 \cap b_5 = 0 & \quad \text{10 10 5 Yukawas} \quad E_6 \text{ enhancement,} \\ b_3^2 - 4b_0 b_4 = 0 \cap b_3 = 0 & \quad \text{10 }\bar{5}\text{ }\bar{5}\text{ Yukawas} \quad SO(12) \text{ enhancement.} \end{aligned} \qquad (5.7)$$

In the $SO(10)$-case we have the following Yukawa couplings:

$$\begin{aligned} b_3 = 0 \cap b_4 = 0 & \quad \text{16 16 10 Yukawas} \quad E_7 \text{ enhancement,} \\ b_2^2 - 4b_0 b_4 = 0 \cap b_3 = 0 & \quad \text{16 10 10 Yukawas} \quad SO(14) \text{ enhancement.} \end{aligned} \qquad (5.8)$$

By constructing the base manifold B and the elliptically fibered Calabi-Yau fourfold we are able to give explicit expressions for the quantities defined above. Furthermore, knowing the homology classes of divisors we can obtain intersection numbers and other topological data of the GUT brane, the matter curves and the Yukawa couplings. In order to make these calculations we make use of toric geometry. In the following subsections, we will explain the necessary ingredients for these computations; see also chapter 2.

5.1.2 Base manifolds

In our work, we have considered toric ambient spaces from normal fans of reflexive polytopes. There are three reasons for this choice. First, these toric varieties have well understood singularity properties. Second, we know how to calculate their Hodge numbers in terms of combinatorial formulas due to the works [28]; see formula (2.37). Third, we have a classification scheme for reflexive polytopes up to dimension four [30].

A toric variety X_Σ is smooth iff all cones of Σ are simplicial and basic (i.e. generated by a subset of the lattice basis). The normal fan of a given reflexive polytope will not generally satisfy these conditions. However, in our setup, we can always resolve singularities in toric spaces by subdivisions of their fan [40, 41, 98]. Take the polytope $\Delta^\circ \subset N$ with all its lattice points, and consider a star triangulation thereof, i.e. a triangulation where the maximal simplices always contain the origin. The fan over the facets of this polytope depends on the particular star triangulation we have chosen. Then reflexivity implies that there are no singularities at codimension lower than four. For a four-dimensional polytope, hence, there can be only point-like singularities. A hypersurface without fixed points can always be deformed to avoid this kind of singularities. Hence, for our setups, a base point free (Cartier) divisor is smooth.

The intersection ring of a non-singular compact toric variety is given by the quotient ring (cfr. formula (2.40))

$$\mathbb{Z}[D_1, \ldots, D_r] / \langle I_{SR}, I_{lin} \rangle . \tag{5.9}$$

Here I_{SR} is the Stanley-Reisner ideal with relations of the type $D_{j_1} \cdot \ldots \cdot D_{j_l} = 0$ for elements of the minimal index set I. Furthermore, one must mod out the ideal I_{lin} generated by the linear relations $\sum_j \langle m, v_j \rangle D_j = 0$. The intersection ring of an embedded hypersurface is given by restricting the intersection ring of the ambient space to the divisor D describing the hypersurface as follows:[1]

$$D_{j_1} \cdot \ldots \cdot D_{j_{n-1}}|_D = \int_D D_{j_1} \wedge \ldots \wedge D_{j_{n-1}} = \int_X D_{j_1} \wedge \ldots \wedge D_{j_{n-1}} \wedge D . \tag{5.10}$$

We need the Kähler cone of the toric variety to determine the volumes of the divisors. With this information we will be able to make statements about the existence of a decoupling limit. We obtain it by starting from its dual, the Mori cone. The Mori cone is the cone of (numerically) effective curves. We determine it using the Oda-Park algorithm [23, 36], that has been implemented in the new version of the PALP code [1, 2].[2] The extended PALP uses the SINGULAR [38] program to determine the intersection ring. In what follows, we approximate the Kähler cone of the embedded hypersurface by that of the ambient space. Since there could be more effective curves on the hypersurface than the induced ones, the Kähler cone of the hypersurface may be smaller than the one of the ambient space.

---
[1] By abuse of notation D denotes the divisor as well as the associated Poincaré dual element of the cohomology.
[2] The Mori cone of the ambient space is computed with option "-m" of the PALP-program `mori.x`; see chapter 3.

Induced divisors

In our setup the base manifold is a divisor embedded in a toric ambient space. The reader may ask under which conditions and to which extent the homology of the hypersurface is induced from the homology classes of the toric ambient space. Indeed, not all toric divisors of the ambient space may induce a divisor on the hypersurface. For a Calabi-Yau hypersurface given by a reflexive polytope Δ°, this is the case if we have a divisor $D_{\text{int}.i}$ obtained from points that lie in the interior of a facet of the polytope. To observe this, we consider the intersection product, on the CY hypersurface, of some $D_{\text{int}.i}$ with divisors not coming from interior points,

$$D_{\text{CY}} \cdot D_{\text{int}.i} \cdot D_{j_1} \cdot \ldots \cdot D_{j_{n-2}} = n_{i\,j_1\ldots j_{n-2}} \,. \tag{5.11}$$

We add to this equation intersection products of the form:

$$D_j \cdot D_{\text{int}.i} \cdot D_{j_1} \cdot \ldots \cdot D_{j_{n-2}} = 0 \,, \tag{5.12}$$

where the D_j is a divisor that does not lie on the facet of the $D_{\text{int}.i}$. This intersection is zero because the fan of the toric space is obtained from a maximal triangulation of the defining lattice polytope. Hence, divisors that lie in the interior of a facet intersect only divisors that also lie on that facet. The lattice polytopes that we consider are reflexive. Thus, for each facet f_i of the polytope we have a point $m_{f_i} \in M$ in the dual lattice polytope with $\langle m_{f_i}, p_j \rangle = -1$ for all points $p_j \in f_i$. From m_{f_i} we obtain the principal divisor

$$D_{m_{f_i}} = \sum_{p_j \in f_i} -D_j + \sum_{p_k \in \Delta^\circ \setminus f_i} \langle m_{f_i}, p_k \rangle D_k \,. \tag{5.13}$$

Since $D_{\text{CY}} = \sum_{p_k \in \Delta^\circ} D_k$, we can add up (5.11) and (5.12) to

$$D_{m_{f_i}} \cdot D_{\text{int}.i} \cdot D_{j_1} \cdot \ldots \cdot D_{j_{n-2}} = -n_{i\,j_1\ldots j_{n-2}} \,. \tag{5.14}$$

A principal divisor always has intersection number zero with any other divisor, hence, we obtain $n_{i\,j_1\ldots j_{n-2}} = 0$. Therefore, the divisor $D_{\text{int}.i}$ does not intersect with the Calabi-Yau hypersurface.

In the case of a hypersurface with a generic (multi) degree, we cannot use the above M-lattice vector to prove that divisors obtained from interior points do not lie on the hypersurface. However, we may find another vector m such that its principal divisor is the sum of the divisor of the hypersurface and the sum of toric divisors that do not come from points of the considered facet.

For the general hypersurface case not only divisors coming from interior points of facets may not induce a divisor but also others. For example, the lower bound on the hypersurface degrees that we will consider below is that they include all homogeneous coordinates. At the bound we may encounter situations where one of the toric divisors has the same weight as the hypersurface. In this case, all toric divisors that do not intersect the divisor showing up linearly in the hypersurface equation will not lie on the hypersurface.

# of points	# of vertices	# of polytopes
6	5	3
7	5	7
7	6	18
8	5	9
8	6	70
8	7	89
9	5	13
9	6	115
9	7	406
9	8	358
		1088

Table 5.1: Lattice polytopes specifying toric ambient spaces for B.

Toric data for base manifolds

In this section, we introduce the class of base manifolds B we will be working with. We will consider base geometries that are non-negatively curved hypersurfaces in a toric ambient space. We restrict to hypersurfaces with hyperplane class positive and strictly smaller than the class of the anti-canonical bundle of the ambient space. An interesting class of manifolds to look at would be Fano threefolds. However, as has been argued in [99], Fanos do not allow for a decoupling limit. We are thus forced to look for more general hypersurfaces. In [81,83,95], such examples have been obtained by constructing point and curve blow-ups of those Fano threefolds which are hypersurfaces in $\mathbb{P}^4$. A systematic construction for up to three point and curve blow-ups has been undertaken in [95] by a classification of the weight systems specifying the toric ambient space. What we would like to achieve here is to construct base manifolds in a more general class of ambient spaces, using toric geometry. In order to do so we will use a slightly different point of view than in [95]: instead of classifying weight systems corresponding to blow-ups we will specify the ambient space by reflexive polyhedra in four dimensions. These have been classified in [30]. Since we are not looking for Calabi-Yaus each of these polytopes will give us a large number of models since there are typically many possibilities to define hypersurfaces inside the ambient space defined by the polytope that fulfill the above above hyperplane class constraint. Therefore it has not been possible for us to construct base manifolds from all the 473 800 776 reflexive polyhedra in four dimensions. Instead, we will look at a class of geometries specified by N-lattice polytopes which define toric ambient spaces that are fourfolds with Picard number less than five. Concretely, we have looked at N-lattice polytopes with up to nine points, including the origin. Not all the points of a polytope are also vertices. We have divided up the data accordingly. This is summarized in table 5.1. The polytope data can be recovered from this information at [34]. The points of the N-lattice polytopes encode the weight matrices which we can recovered using PALP. The next step in constructing the base manifolds is to specify a hypersurface of degrees d_i, where i runs over the rows in the weight matrix. The type of hypersurface we are interested in constrains the number of possible degrees. If $d_i = \sum_j w_{i,j}$, where w_{ij} are the homogeneous weights of the variables, the hypersurface will be Calabi-Yau. This gives an upper bound for the degrees: for our purposes we have to consider hypersurface degrees such that at least one of the d_i is strictly smaller than the sum of the weights. Furthermore, we would like our base manifold B to be a genuine complex codimension one hypersurface inside the toric ambient space. Therefore, we impose the condition that each variable has to appear in at least one monomial of the hypersurface equation. If the homogeneous weight of a variable is higher than the hypersurface degree the variable will certainly not appear in the hypersurface equation. This gives a lower bound on the hypersurface degree. Since this bound is necessary but not sufficient, one has to check for each model if indeed all the variables appear in the hypersurface equation. For the ambient spaces specified by the 1088 polytopes above we have constructed all the hypersurfaces satisfying these conditions. In this way we have obtained as many as 569 674 potential candidates for bases of an F-theory compactification.

GUT data from base manifolds

Even though we are ultimately interested in constructing a full F-theory compactification on a Calabi-Yau fourfold, a lot of important information about the GUT model is already encoded in the geometry of the base manifold. What is more, in many cases this data can be inferred from the toric data of the ambient space. In the following we discuss what we can learn from the geometry of B and how to compute phenomenologically relevant data using toric geometry. In our discussion about the GUT brane S, which wraps a toric divisor in B, we will focus on $SU(5)$ and $SO(10)$ models.

Singularities

Singularities can either come from singularities of the ambient space or the hypersurface equation. Since the ambient space of the base manifold is characterized by a reflexive polytope in four dimensions, only point-like singularities arise there. On the other hand the hypersurface itself can be singular. A hypersurface given by an equation $W(x_1, \ldots x_n) = 0$ is singular at a locus x_{sing} if:

$$W|_{x_{sing}} = 0 \quad \text{and} \quad \partial_{x_i} W|_{x_{sing}} = 0, \qquad x_{sing} \in X_6 \quad i = 1, \ldots, N. \tag{5.15}$$

A sufficient condition for regularity is that the divisor defining the hypersurface is base point free. In this case the hypersurface can be transversally deformed in every point. By Bertini's theorem, it will not have any singularities of the kind of (5.15). Additionally, the hypersurface will miss possible point singularities of the ambient space which are the only singularities of our toric ambient spaces of B. The base point free condition is given purely in terms of the combinatorics of the lattice polytope and therefore quite simple to check; see section 2.2.2.

Almost Fano manifolds

An almost Fano threefold is an algebraic threefold that has a non-trivial anti-canonical bundle with at least one non-zero section at every point. Our toric construction of base manifolds does not necessarily lead to almost Fano manifolds. Thus, we check this criterion by explicitly searching for non-zero sections in every example. In the examples analyzed in [95], a connection between the almost Fano property of B and the reflexivity of the lattice polytope associated to the elliptically fibered fourfold had been observed.

Del Pezzo divisors

Having specified a base manifold B, the next task is to identify suitable GUT divisors S. For this purpose, we will systematically search for del Pezzo divisors inside B. There are several motivations to look for del Pezzos. In local F-theory GUTs, the del Pezzo property ensures the existence of a decoupling limit [77, 78]. For $SU(5)$ GUT models, the fact that del Pezzos have $h^{0,1} = h^{2,0} = 0$ implies some powerful vanishing theorems which forbid exotic matter after breaking $SU(5)$ to the Standard Model gauge group [78]. However, one should keep in mind that there are other possibilities besides del Pezzos: as pointed out in [76], for the F-theory model to have a heterotic dual, S may also be a Hirzebruch or an Enriques surface. Recently, a construction of an F-theory GUT on an Enriques surface has been discussed [100].

We will identify candidates for del Pezzo divisors inside B by their topological data. All the calculations can be done using toric geometry. Suppose the base manifold has hyperplane class which, by abuse of notation, we also call B and is embedded in a toric ambient space with toric divisors D_i. The total Chern class of a particular divisor S in B is

$$c(S) = \frac{\prod_i (1 + D_i)}{(1 + B)(1 + S)}. \qquad (5.16)$$

A necessary condition for the divisor S to be dP_n is that it must have the following topological data:

$$\int_S c_1(S)^2 = 9 - n, \quad \int_S c_2(S) = n + 3 \quad \Longrightarrow \quad \chi_h = \int_S \mathrm{Td}(S) = 1, \qquad (5.17)$$

where χ_h is the holomorphic Euler characteristic. Since del Pezzos are Fano twofolds, we have a further necessary condition. The integrals of $c_1(S)$ over all torically induced curves[3] on S have to be positive:

$$D_i \cap S \cap c_1(S) > 0, \quad D_i \neq S, \quad \forall D_i \cap S \neq \emptyset. \qquad (5.18)$$

Genus of matter curves

Assuming that we have set up the right GUT theory on the divisor S, matter is localized at curves of further enhancement of the singularity. The curve classes M of the matter curves

---
[3] Of course, positivity should hold for all curves, but within the framework of toric geometry we can only verify this for the divisors induced from the ambient space.

can be expressed in terms of the toric divisors of the ambient space. The genus of the matter curve can be computed using its first Chern class and the triple intersection numbers. The total Chern class is

$$c(M) = \frac{\prod_i (1+D_i)}{(1+B)(1+S)(1+M)}. \tag{5.19}$$

After expanding this expression to obtain $c_1(M)$, the Euler number can be calculated by the following intersection product:

$$\chi(M) = 2 - 2g(M) = c_1(M) \cap M \cap S. \tag{5.20}$$

Note that we have made the assumption that the matter curves are generic and do not factorize. This may not always be the case and then formula (5.20) will yield the sum of the Euler numbers of the factorized curves as result. This may for instance lead to negative values for the genus of the matter curve if we naïvely assume a single connected curve. The genus of M gives us information about the number of moduli on the matter curve. Since these moduli will eventually have to be stabilized, matter curves of low genus are desirable from a phenomenological point of view.

Yukawa points

Yukawa couplings arise at points inside B where the GUT singularity has a rank 2 enhancement. In the generic situation the equations specifying the Yukawa points can be expressed as classes Y_1, Y_2 in terms of the toric divisors. The number of Yukawa points is then given by the following intersection product:

$$n_{\text{Yukawa}} = S \cap Y_1 \cap Y_2. \tag{5.21}$$

In order to account for the Standard Model Yukawa couplings, only a small number of Yukawa points is needed. In $SO(10)$ models, for example, all the Standard Model couplings descend from **16 16 10** Yukawas, which is why it would be nice to find a geometry where the number of **16 10 10** Yukawa points is as small as possible. Most of the known global geometries come with a large number of Yukawa points. The situation is particularly bad for dP_n with small n [101]. Our analysis shows however that dP_0 and dP_1 are by far the most common del Pezzo divisors in the base manifolds.

Decoupling limit

One of the key issues which allows for the discussion of GUT models within F-theory locally around the 7-branes is the existence of a decoupling limit. The Planck mass and the mass scale of the GUT theory are related to the geometry in the following way:

$$M_{pl}^2 \sim \frac{M_s^8}{g_s^2} \text{Vol}(B), \qquad M_{GUT} \sim \text{Vol}(S)^{-\frac{1}{4}}, \qquad 1/g_{YM}^2 \sim \frac{M_s^4}{g_s} \text{Vol}(S), \tag{5.22}$$

see for instance [102]. Therefore one has:

$$\frac{M_{GUT}}{M_{pl}} \sim g_{YM}^2 \frac{\text{Vol}(S)^{3/4}}{\text{Vol}(B)^{1/2}}. \tag{5.23}$$

There are two ways to achieve a small value for M_{GUT}/M_{pl}. These are often referred to as the physical and the mathematical decoupling limit. In the physical decoupling limit, the volume of the GUT brane S is kept finite while $\text{Vol}(B) \to \infty$. The mathematical decoupling limit takes $\text{Vol}(S) \to 0$ for finite volume of B. In the case of a rigid del Pezzo divisor, the mathematical decoupling limit should always be possible. Thus, it can be used to check whether a del Pezzo is rigid. Here we study the dependence of the volumes of S and B in terms of the Kähler moduli. This discussion tells us if a decoupling limit can in principle be realized in the given geometry. If the limits are actually realized is a question of moduli stabilization, which we will not discuss here.

The question of whether there exists a decoupling limit can again be addressed within the realm of toric geometry. In order to obtain positive volumes, we must find a basis of the Kähler cone. The Kähler cone of the hypersurface describing the base is hard to compute. Therefore we will approximate it by the Kähler cone of the ambient space. Having found a basis K_i of the Kähler cone, the Kähler form J can be written as $J = \sum_i r_i K_i$ with $r_i > 0$. Using the Mori cone we can express K_i in terms of the toric divisors D_i. The triple intersection numbers restricted to B allow us to compute the following volumes in terms of the Kähler parameters r_i:

$$\text{Vol}(B) = J^3, \qquad \text{Vol}(S) = S \cdot J^2. \tag{5.24}$$

The existence of a mathematical and physical decoupling limits can be deduced from the moduli dependence of these volumes. As was first observed in [83], these two decoupling limits may be governed by different vectors in the Kähler cone.

5.1.3 Elliptically fibered Calabi-Yau fourfolds

Construction of the fourfolds

We now go on to construct an elliptically fibered Calabi-Yau fourfold from B. We obtain such an elliptic fibration by first fibering $\mathbb{P}_{231}[6]$ over the toric ambient space of the base manifold. Thus, we extend the weight matrices describing the ambient space of B by suitable weights for the new fiber coordinates (x, y, z). This is done such a way that x, y, and z transform as K_B^{-2}, K_B^{-3}, and $\mathcal{O}_B$, respectively. We also add an extra weight vector $(2, 3, 1, 0, 0, \ldots, 0)$ to account for the $\mathbb{P}_{231}$. In order to have a well defined torus fibration, the coefficients a_i of equation (5.2) have to be sections of K_B^{-n} with some appropriate power n. The sums of the degrees of the hypersurface equation of the base and of the equation specifying the elliptic fibration are now equal to the degree of the anti-canonical bundle of the ambient toric sixfold. Hence, the complete intersection of these two equations is a Calabi-Yau manifold. This variety may be singular in some cases. The complete intersection Calabi-Yaus we consider here are given in terms of a pair of reflexive lattice polytopes Δ and Δ°, together with a nef-partition:

$$\Delta = \Delta_1 + \ldots + \Delta_r \qquad \Delta^\circ = \langle \nabla_1, \ldots, \nabla_r \rangle_{\text{conv}}$$
$$(\nabla_n, \Delta_m) \geq -\delta_{nm} \qquad (5.25)$$
$$\nabla^\circ = \langle \Delta_1, \ldots, \Delta_r \rangle_{\text{conv}} \qquad \nabla = \nabla_1 + \ldots + \nabla_r$$

Here, $\langle \ldots \rangle_{\text{conv}}$ denotes a convex hull of lattice polytopes, and $\Delta = \Delta_1 + \ldots + \Delta_r$ (and analogously for ∇) is a Minkowski sum.

The extension of the weight systems of the base threefold is straightforward. However, there are several issues of both conceptual and technical nature which prevent us from constructing an F-theory compactification for every base B. These are discussed in the following.

Software constraints

There are two main constraints affecting our search for complete intersection Calabi-Yaus (CICYs). First, PALP was originally designed to analyze complete intersection Calabi-Yaus of the type (5.25), which does not cover all the possibilities we encounter in our construction of global F-theory GUTs. The software efficiently analyzes combined weight systems to find their description in terms of (six-dimensional) reflexive polytopes. Afterwards PALP determines their nef-partitions and the Hodge numbers of the CICY. Given a six-dimensional reflexive polytope describing the ambient space, the common zero locus of any two transversal equations is a suitable Calabi-Yau. Note that the two defining equations do not have to descend from the nef-partitions, but only for nef partitions it is known how to determine the Hodge numbers of the CICY in terms of combinatorial data [103]. Thus, we could only do detailed calculations for examples that fulfill the requirements of (5.25). In fact, not all of the combined weight systems we have constructed extending the base weight matrices correspond to reflexive polytopes or do have nef partitions. Table 5.5 in section 5.2 shows how many CICYs satisfy these conditions. Reflexivity has turned out to be a severe constraint.

The second obstacle in our analysis of the fourfolds is that due to computational constraints we have not been able to determine the six-dimensional N-polytopes for all weight matrices. The last column of table 5.5 shows where the software has failed. The entries in the columns give information of two types of errors that can occur when determining the polytopes in the N-lattice: in most cases the error comes from the the issue that PALP cannot determine the N-lattice polytope by solving the equations encoded in the weight matrices. This problem might in principle be overcome by choosing the points of the N-lattice polytope as an input instead of the weight matrix. In fewer cases the N-lattice polytope can be found but an upper bound to the number of points is violated. The upper bound could be increased but that usually leads to very long computation times. The error distribution is in agreement with the intuitive idea that the complexity of the weight matrices increases with the number of points. For the fibrations over polytopes with 8 points where 7 of which are vertices we get an error in $10,9\%$ of the cases, for polytopes with 9 points and 8 vertices we have an error occurrence of $28,5\%$.

The fourfold data available at [104] do not contain the Hodge numbers of the CICYs. They can be easily determined with help of the nef-function of PALP.[4] However, due to the complexity of the polytopes their calculation would have been too time consuming to be applied to every model we had.

Compatibility with the elliptic fibration

Once we have found a Calabi-Yau fourfold characterized by a pair of dual polyhedra and its nef-partitions, we still need to make sure that one of the nef-partitions is compatible with the desired elliptic fibration. The most elementary requirement for a well-defined Weierstrass model is of course that the points in Δ° corresponding to the coordinates of the torus fiber are all in the same component of the nef-partition. However, this criterion is not sufficient in order to recover the desired Weierstrass model. We also have to make sure that the coefficients a_n in (5.2) transform appropriately as sections of K_B^{-n}. This translates into conditions on the on the (sums of) weights of the variables in the individual nef-partitions.

Engineering GUT models

By now we have constructed complete intersection Calabi-Yau fourfolds of type (5.1). The next step is to obtain a GUT model. This is achieved by imposing the factorization constraints such as (5.3) or (5.4) on the coefficients $a_r(y_i, w)$ in the Tate equation (5.2). The procedure can be done within the toric framework, as has been proposed first in [81]. The hypersurface constraints can be recovered from the toric data as follows:

$$f_m = \sum_{w_k \in \Delta_m} c_k^m \prod_{n=1}^{2} \prod_{\nu_i \in \nabla_n} x_i^{\langle \nu_i, w_k \rangle + \delta_{mn}} \qquad m, n = 1, 2, \tag{5.26}$$

where the c_k^m are complex structure parameters. The Tate form (5.2) implies that the a_n appear in the monomials which contain z^n. We can isolate these monomials by identifying the vertex ν_z in (∇_1, ∇_2) that corresponds to the z-coordinate. All the monomials that contain z^r are then in the following set:

$$A_r = \{w_k \in \Delta_m : \langle \nu_z, w_k \rangle - 1 = r\} \qquad \nu_z \in \nabla_m, \tag{5.27}$$

where Δ_m is the dual of ∇_m, which denotes the polytope containing the z-vertex. The polynomials a_r are then given by the following expressions:

$$a_r = \sum_{w_k \in A_r} c_k^m \prod_{n=1}^{2} \prod_{\nu_i \in \nabla_n} y_i^{\langle \nu_i, w_k \rangle + \delta_{mn}} \Big|_{x=y=z=1}. \tag{5.28}$$

Now we can remove all the monomials in a_r that do not satisfy the factorization constraints of the singularity classification. In order to perform this calculation, we have to identify the fiber coordinates (x, y, z) and the GUT coordinate w within the weight matrix of the fourfold.

---

[4]In fact **nef.x** yields the Hodge numbers by default. The flag "-p" deactivates their calculation. For more details, we refer to the help information: **nef.x -h**.

The restriction to a specific GUT group amounts to removing a considerable amount of M-lattice points. As has been observed in [95], these manipulations may destroy the reflexivity of the polytope. The dual polytope in the N-lattice will have acquired additional points that can be interpreted as exceptional divisors obtained by blowing-up the GUT singularity [81,83].

$U(1)$-restricted models

Recently there has been active discussion in the literature on how to globally define fluxes in F-theory models. While a full answer to this problem is still unknown there has been some progress in incorporating the spectral cover construction into global models [87,88]. For phenomenological reasons, one has to make sure that, in $SU(5)$ models, the spectral cover splits. This is necessary to forbid dimension four proton decay operators. In $SO(10)$ models a split spectral cover is used to generate chiral fermions [95, 105]. However, as has been argued in [86,87] the local picture of a split spectral cover may in general not be sufficient. The authors of [87] have shown that a lift of the local split spectral cover construction to a globally defined "$U(1)$-restricted Tate model" can give the needed further selection rule. This is achieved by imposing a global $U(1)_X$-symmetry in the elliptic fibration. In terms of the Tate model, this is achieved by setting $a_6 = 0$. In the toric language this corresponds to removing even more points in the M-lattice, in addition to the manipulations needed for imposing the GUT model. Due to this procedure the Euler number decreases significantly, which is problematic for tadpole cancellation. Since the $U(1)$-restriction removes even more points from the M-lattice, reflexivity might not be maintained.

5.2 Data analysis

In this section, we analyze our data.[5] We have produced 569 674 base geometries in total. We will discuss their properties and the associated elliptically fibered fourfolds.

5.2.1 Base manifolds

We collect the information about the base geometries in several tables. Our discussion will be concerned with properties of the base manifold, properties of its divisors and furthermore matter curves, Yukawa couplings as well as the existence of a decoupling limit.

In table 5.2, we summarize some information about the base geometries. We subdivide the models into classes pnvm, denoting models based on polytopes which have n points and m vertices. The last three columns in the table indicate how many of the base manifolds are Cartier divisors, base point free or almost Fano. We note that base point freedom and in particular the almost Fano property are extremely rare items. As for almost Fano, it turns out

---

[5]The complete data concerning the base manifolds, their analysis, as well as the elliptically fibered fourfolds and the GUT models is available at [104]. For details on the data format, we refer to the README.txt file the reader can find there.

class	# of polytopes	# of base manifolds	Cartier	BP-free	almost Fano
p6v5	3	12	6	6	10
p7v5	7	155	66	31	39
p7v6	18	307	199	131	94
p8v5	9	812	424	86	73
p8v6	70	6691	3265	816	584
p8v7	89	8168	4464	1542	779
p9v5	13	8238	1243	77	155
p9v6	115	84848	27037	1651	1542
p9v7	406	257024	107119	10515	5955
p9v8	358	203419	101562	14564	5677
total	1088	569674	245385	29401	14908

Table 5.2: Analysis of the base manifolds.

that this property of the base manifold is not needed in order to have a Calabi-Yau fourfold that is characterized by a reflexive polytope.

In our search for geometries that are suitable for F-theory model building, we have focused on identifying del Pezzo divisors inside the base manifold. The results of our search are summarized in table 5.3. All the divisors in this counting satisfy (5.17) and (5.18). Among all the base geometries, we have identified 269 636 models with del Pezzo divisors, and a total number of 471 844 del Pezzos. The dP_n with $n = 0, 1, 2$ are the most common ones. So far, our discussion has included all possible choices of base manifolds. We can now collect those models that have some attractive features. For that reason, we will now focus on those models where B is regular and has at least one del Pezzo divisor that allows for a mathematical or physical decoupling limit. This leaves us with only a small fraction of models, as indicated in table 5.4. In the first column, we count the number of models where the hypersurface divisor of B is Cartier and there is at least one del Pezzo divisor with a mathematical or physical decoupling limit. In the second column, we furthermore implement the constraint that B is base point free. In the third column, we count the total number of all del Pezzos (also those without decoupling limit) in the base point free geometries, where at least one dP-divisor allows for a decoupling limit.

5.2.2 Fourfolds

In this section, we discuss the Calabi-Yau fourfolds which are elliptic fibrations over the base threefolds. The toric data of the fourfolds is obtained by extending the weight matrices associated to the base manifolds, as discussed in section 5.1.3. Complete intersection Calabi-Yaus can be analyzed by PALP. The fourfold data contains a lot of information, which is relevant for finding global F-theory GUT models. We can use the data to answer the following questions:

class	base manifolds with dPs	# of dP_n	dP_0	dP_1	dP_2	dP_3	dP_4	dP_5	dP_6	dP_7	dP_8
p6v5	6	25	9	6	-	-	-	-	6	4	-
p7v5	66	150	36	72	4	-	2	1	17	14	4
p7v6	206	597	121	239	35	11	17	9	73	64	28
p8v5	429	787	133	431	43	-	14	8	75	45	38
p8v6	3322	6259	1074	2883	539	157	164	171	520	458	293
p8v7	4888	11449	1868	4162	1325	670	451	532	931	947	563
p9v5	3213	5415	1562	1740	274	61	115	31	617	949	66
p9v6	31160	45039	8598	20261	4228	1167	992	1023	3763	3823	1184
p9v7	113364	181672	31926	72056	22238	9432	5812	6632	12061	13839	7376
p9v8	112982	220451	35669	73549	32191	18130	11098	11394	14950	15183	8887
total	269636	471844	80996	175399	60877	29628	18665	19801	33013	35326	18439

Table 5.3: Results of the del Pezzo analysis.

class	Cartier+dec+dP	BP-free+dec+dP	# dP for BP-free+dec
p6v5	-	-	-
p7v5	29	22	74
p7v6	85	72	277
p8v5	224	74	212
p8v6	1492	665	2073
p8v7	2412	1264	4490
p9v5	726	62	239
p9v6	10900	1332	3334
p9v7	46142	8933	26776
p9v8	53356	13108	50930
total	115366	25532	88405

Table 5.4: Base manifolds with del Pezzo divisors and decoupling limit.

1. Does the extension of the weight matrix of the base lead to a reflexive polytope?

2. How many of the Calabi-Yau fourfolds have nef-partitions that are compatible with the elliptic fibration over B?

3. Do the "good" base manifolds (i.e. those which are regular, have del Pezzo divisors and a decoupling limit) always extend to Calabi-Yau fourfolds, which are described in terms of reflexive polytopes and nef-partitions?

4. After imposing a GUT group using the construction of [81], are the fourfold polytopes still reflexive?

5. Does imposing the GUT model lead to further non-abelian enhancements on divisors other than the GUT divisor?

6. Can we implement a $U(1)$-restricted Tate model in order to impose a global $U(1)$-symmetry [87] without destroying desirable properties on the Calabi-Yau fourfold?

Even though we have the tools to answer all these questions, working out the details for a large class of models is quite tricky and takes up a lot of computing time. This is why we will address some of these issues, in particular the fifth question, only in several examples.

We start by answering the first question above. As a somewhat surprising outcome, only a very small fraction of threefold base manifolds can be extended to a Calabi-Yau fourfold which is described by a pair of reflexive polyhedra and at least one nef-partition. We have found 27 345 such models. The results are summarized in table 5.5. About one quarter of the extended weight systems could not be analyzed due to their complexity.

For the rest of the discussion we will focus on those fourfolds which can be characterized by reflexive polytopes and have at least one nef-partition. At first we merge the fourfold

class (base)	reflexive+nef part.	reflexive, no nef part.	non-reflexive	PALP errors
p6v5	10	-	2	-
p7v5	65	6	84	-
p7v6	128	7	172	-
p8v5	197	103	308	$188 + 16$
p8v6	1170	344	4481	$660 + 36$
p8v7	1051	267	5958	$892 + 0$
p9v5	256	146	583	$7187 + 66$
p9v6	4033	3530	61211	$14861 + 1213$
p9v7	12101	8963	176598	$58439 + 928$
p9v8	8334	5266	131835	$57918 + 66$
total	27345	18632	381232	$140145 + 2325$

Table 5.5: Fourfold polytopes.

class	CY_4+refl+nef	Cartier base+dP_n+dec.	BP-free base+dP_n+dec.
p6v5	10	-	-
p7v5	65	24	18
p7v6	128	61	57
p8v5	197	94	38
p8v6	1170	685	402
p8v7	1051	760	591
p9v5	256	5	-
p9v6	4033	1679	414
p9v7	12101	6909	2714
p9v8	8334	5794	3152
total	27345	16011	7386

Table 5.6: CY fourfolds where the base manifolds are suitable for F-theory model building.

data with the data of the base manifold in order to check how many of the "good" base manifolds also lead to Calabi-Yau fourfolds that are characterized by reflexive polytopes with nef-partitions. Our findings are collected in table 5.6. The number of models which have a reflexive fourfold polytope, where the base is regular and there is at least one del Pezzo divisor with a mathematical and/or physical decoupling limit, is 7386. In table 5.7, we list the distribution of del Pezzos in these "good" models. Even if we have a reflexive fourfold polytope with nef-partitions it is not implied that the nef-partitions are compatible with the elliptic fibration over B. The extended weight systems will always lead to elliptic fibrations, but not necessarily over the base manifold we want. In many cases, there may even be more than one nef-partition that is compatible with the elliptic fibration over B. However, these nef-partitions always lead to the same Tate model. Taking this into account, we are left with 3978 Calabi-Yau fourfolds. Our results can be found in table 5.8. With a nef-partition in hand,

class	# models	# of dP_n	dP_0	dP_1	dP_2	dP_3	dP_4	dP_5	dP_6	dP_7	dP_8
p6v5	-	-	-	-	-	-	-	-	-	-	-
p7v5	18	66	17	39	-	-	2	-	6	2	-
p7v6	57	212	39	92	10	7	7	5	26	21	5
p8v5	38	100	6	86	-	-	3	-	2	3	-
p8v6	402	1198	172	696	83	44	48	39	42	68	6
p8v7	591	2287	284	894	287	192	124	154	131	178	43
p9v5	-	-	-	-	-	-	-	-	-	-	-
p9v6	414	855	102	494	91	44	33	27	30	30	4
p9v7	2714	7378	902	3383	1122	931	375	384	198	324	59
p9v8	3152	12334	1377	4161	2343	1605	768	881	533	507	159
total	7386	24430	2899	9845	3936	2823	1360	1490	968	1133	276

Table 5.7: Distribution of del Pezzos in "good" F-theory geometries.

class (base)	# of models w/ ell. comp. nef	# of ell. comp. nef
p6v5	-	-
p7v5	4	6
p7v6	46	83
p8v5	3	5
p8v6	110	215
p8v7	445	1157
p9v5	-	-
p9v6	69	116
p9v7	1014	2538
p9v8	2287	7677
total	3978	11797

Table 5.8: CY fourfolds with Tate models.

we can go on to construct GUT models for a particular gauge group, as described in section 5.1.3. For the 3978 fourfold geometries in table 5.8 that have a nef-partition that is compatible with the elliptic fibration, we have constructed $SU(5)$ and $SO(10)$ GUT models on every del Pezzo divisor. In order to make this calculation, we have to identify the coordinates of the torus fiber and the GUT divisor in the toric data of the Calabi-Yau fourfold. This can be done by matching the columns of the weight matrix of B with the columns of the weight matrix of X_4. Note that this identification may not always be unique due to symmetries of the weight matrix. Of course, the different choices do not lead to different GUT models. One prominent example of a weight matrix with such a symmetry is the dP_5-model discussed in [83].

Carrying out this procedure we get a total number of 45 304 global F-theory GUTs. After removing redundancies coming from symmetries in the weight matrix, we are still left with 30 922 models. Note however that not all of these models will be usable, since the removal of

	with redundancies			without redundancies		
type	refl.	non-refl.	no nef	refl.	non-refl.	no nef
$SU(5)$	17099	5553	-	11275	4186	-
$SO(10)$	16625	6020	7	10832	4622	7
$SU(5)+U(1)$-restr.	17099	5553	-	11275	4186	-
$SO(10)+U(1)$-restr.	16625	6020	7	10832	4622	7

Table 5.9: Reflexivity of polytopes after implementing the GUT group.

points in the M-lattice in order to implement the GUT group may destroy the reflexivity of the polytope. In very few examples, it might also happen that there is no longer a nef-partition. We collect this information in table 5.9. We make two observations: first, in about one third of the models, imposing the GUT group destroys reflexivity, and second, the $U(1)$-restriction, does not put any further constraints on the reflexivity of the polytopes.

In the final step of our data analysis, we search for new examples of F-theory GUTs that might be interesting for string phenomenology. Therefore we would like to isolate models where the GUT divisor S has matter curves with a small number of moduli and not too many Yukawa points. Even though the geometries we have started with have GUT divisors with very diverse topological data, the cuts we have imposed put severe restrictions on the geometry and as a consequence also on the topological numbers of the divisors. In table B.1 in the appendix, we list the matter genera and Yukawa points for $SU(5)$ and $SO(10)$ del Pezzos with a physical decoupling limit, and their occurrence in global models where the fourfold polytopes are reflexive after imposing the GUT group with or without $U(1)$-restriction. Similar results can be obtained for del Pezzos with a mathematical decoupling limit.

5.2.3 Examples

We will now discuss some examples in more detail. We focus mostly on dP_7 and dP_8 since they are quite rare and dP_8's have not been discussed previously in the context of global models. We will also make some comments on the calculation of Euler numbers using the following formula proposed in [81]: given a resolved Calabi-Yau fourfold with GUT group G, denoted by $\bar{X}_G$, the Euler number is given by

$$\chi_{\bar{X}_G} = \chi_{\bar{X}_4} - \chi_{E_8} + \chi_H, \qquad (5.29)$$

where $\chi_{\bar{X}_4}$ is the Euler characteristic of the resolved X_4 and χ_H denotes a correction related to H, which is the commutant subgroup of G in E_8. The Euler number for a smooth elliptically fibered Calabi-Yau fourfold is

$$\chi_{\bar{X}_4} = 360 \int_B c_1^3(B) + 12 \int_B c_1(B) c_2(B). \qquad (5.30)$$

Defining $\eta = 6c_1(S) + c_1(N_S)$, the correction for $H = SU(n)$ ($n \leq 5$) is given by

$$\chi_{SU(n)} = \int_S c_1^2(S)(n^3 - n) + 3n\eta(\eta - nc_1(S)). \qquad (5.31)$$

Originally, the formula (5.29) was motivated from heterotic/F-theory duality and the spectral cover construction. In [87], (5.29) has been shown to be consistent with mirror symmetry, under which G and H are exchanged. Note that (5.29) is only valid if there are no further non-abelian gauge enhancements away from the GUT brane S. Furthermore, equation (5.29) is not valid for $U(1)$-restricted models. In the following examples, we will see that such extra enhancements can occur and lead to discrepancies in the Euler numbers of the Calabi-Yau fourfolds computed by (5.29) and those Euler numbers obtained by PALP, which uses a formula of Batyrev and Borisov [103].

Three dP_8s

Models where the GUT divisor is a dP_8 are interesting for phenomenology since the genera of the matter curves and the number of Yukawa points are typically low. Unfortunately dP_8s are quite rare in the geometries we have constructed, and it turns out that those appearing in suitable Calabi-Yau fourfolds do not satisfy all the properties we would like to have. We will now discuss three examples.

The base geometry of the first example is encoded in the following weight matrix:

	y_1	y_2	y_3	y_4	y_5	y_6	y_7	y_8	$\sum$	deg
w_1	3	2	1	1	0	1	0	0	8	6
w_2	3	1	1	1	0	0	0	1	7	6
w_3	3	0	1	1	1	0	0	0	6	6
w_4	1	0	0	0	0	0	1	0	2	2

(5.32)

The second but last column indicates the sum of the weights, the last column shows the degrees of the hypersurface equation describing the base manifold B. In our database [104], this model is labeled by (cy4)p9v6n058d6-6-6-2t1. Let us first discuss the properties of B. B is an almost Fano manifold and it is a Cartier divisor that is base point free. Furthermore, we only obtain three induced Kähler classes from the ambient space, since D_7 does not intersect the hypersurface, cf. 5.1.2. There is only one del Pezzo divisor, defined by $y_6 = 0$, which will be our GUT divisor S. The topological data indicates that it is a dP_8. The volumes in terms of Kähler parameters $r_i > 0$ are:

$$\begin{aligned}
\text{Vol}(B) &= 6r_1 r_2^2 + 2r_2^3 + 36r_1 r_2 r_3 + 18r_2^2 r_3 + 54r_1 r_3^2 + 54r_2 r_3^2 + 27r_3^3 + 36r_1 r_2 r_4 + 18r_2^2 r_4 \\
&\quad + 108 r_1 r_3 r_4 + 108 r_2 r_3 r_4 + 162 r_3^2 r_4 + 54 r_1 r_4^2 + 54 r_2 r_4^2 + 162 r_3 r_4^2 + 54 r_4^3, \\
\text{Vol}(S) &= 9 r_3^2.
\end{aligned}$$

(5.33)

It is easy to check that there is a mathematical as well as a physical decoupling limit. Under the mathematical decoupling limit $r_3 \to 0$, S is the only divisor that shrinks to zero size. If we choose $r_1 \to \infty$ as a physical decoupling limit also the divisors $y_2 = 0$ and $y_8 = 0$ remain of finite size. However, studying this base geometry in more detail we see that it is a K3 fibration over $\mathbb{P}^1$. The K3 fiber degenerates at the point, $y_6 = 0$, of the $\mathbb{P}^1$ to a dP_8. Hence, it is a rigid divisor. Constructing a torus fibration over B, we observe that the coefficients a_i of the fibration only depend on the coordinates of the $\mathbb{P}^1$. Thus, the elliptic curve remains constant over the fiber, therefore, also in the case of a degeneration. From the discriminant, we find that the torus degenerates over twelve points of the $\mathbb{P}^1$. Hence, we obtain twelve disconnected branes along the fibers at these points and not a single connected one, as one would expect in the case of a generic fibration.

We can now naïvely proceed and calculate the genera of the matter curves and the Yukawa numbers for a $SU(5)$ GUT on S. We obtain the following:

$$g_{SU(6)} = 11, \quad g_{SO(10)} = 1, \quad n_{E_6} = 0, \quad n_{SO(12)} = 0. \tag{5.34}$$

Due to the absence of Yukawa couplings, this dP_8 is not a good candidate for a viable $SU(5)$ GUT model. However, it still can be used for an $SO(10)$ GUT where the data is as follows:

$$g_{SO(12)} = 2, \quad g_{E_6} = 1, \quad n_{E_7} = 2, \quad n_{SO(14)} = 12. \tag{5.35}$$

The weight matrix (5.32) can be extended to a weight matrix describing a complete intersection Calabi-Yau fourfold X_4. The corresponding six-dimensional lattice polytope is reflexive, and there is one nef-partition that respects the elliptic fibration over B. Using PALP, we can compute the Euler number χ and the non-trivial Hodge numbers for X_4 and for the geometries one obtains after imposing the $SO(10)$ gauge groups. The results are collected in the following table:

type	$h^{1,1}$	$h^{2,1}$	$h^{3,1}$	χ
Tate	12	26	54	288
$SO(10)$	17	29	49	270

(5.36)

As noticed above, already the generic fibration is rather restricted. Thus, we do not obtain 4 for $h^{1,1}$ in the unconstrained case but 12 instead. This indicates that also the $SO(10)$ results should considered with care.

For the $SO(10)$ model, we can compare the Euler number to the result obtained from (5.29), which yields 168. The mismatch implies that some conditions for the validity of this formula are violated. Indeed, looking at the $SU(5)/SO(10)$ Weierstrass model, we find that after imposing the GUT group on the divisor $y_6 = 0$, we also obtain a non-abelian enhancement on the divisor $y_8 = 0$. Comparing with the Tate classification, we get an I_3^s-enhancement for $SU(5)$ on $y_6 = 0$ and an $SU(3)$-enhancement for $SO(10)$. Furthermore, note that removing all the monomials in the Weierstrass equation that do not comply with $SU(5)/SO(10)$, the $(a_0, a_1, a_2, a_3, a_6)$ schematically (i.e. after setting all complex structure parameters to 1) vanish as follows on S: $(1 + w^2, w^2 + w^4, w^2 + w^4 + w^6, w^4 + w^6 + w^8, w^6 + w^8 + \ldots)$ for $SU(5)$, and $(w^2, w^2 + w^4, w^2 + w^4 + w^6, w^4 + w^6 + w^8, w^6 + w^8 + \ldots)$ for $SO(10)$. Thus, the singularity enhancements are actually higher than that of $SU(5)$ or $SO(10)$. As we observed already above, the reason for all the problems roots in the very non-generic form of the coefficients in the Weierstrass model. This comes from the fact that the anti-canonical class does not depend on all toric classes. We see that constructing a Tate model over a promising base manifold may not lead to the wanted brane setup.

As indicated in table B.1, the dP_8 with the matter genera and Yukawa numbers above is the only one with a physical decoupling limit. The dP_8s we have found in the global models we have constructed only have very few combinations of topological numbers. In order to also give an example where an $SU(5)$ GUT is possible, we consider the following base geometry:

	y_1	y_2	y_3	y_4	y_5	y_6	y_7	y_8	$\sum$	deg
w_1	1	1	0	0	0	0	0	0	2	1
w_2	1	0	1	0	1	0	1	0	4	3
w_3	1	0	1	0	0	1	0	1	4	3
w_4	0	0	1	1	1	0	0	0	3	2

(5.37)

The file name in the database is (cy4)p9v8n224d1-3-3-2t1. As in the previous examples the base B is almost Fano. The hypersurface divisor is Cartier and base point free. There are two del Pezzo divisors, one dP_8 and one dP_5. We focus on the dP_8 here, which is given by $y_1 = 0$. The volumes of B and S are:

$$\begin{aligned}
\mathrm{Vol}(B) &= 2r_1^3 + 15r_1^2 r_2 + 6r_1 r_2^2 + 18r_1^2 r_3 + 30r_1 r_2 r_3 + 6r_2^2 r_3 + 18r_1 r_3^2 + 15r_2 r_3^2 + 6r_3^3 \\
&\quad + 18r_1^2 r_4 + 30r_1 r_2 r_4 + 6r_2^2 r_4 + 48r_1 r_3 r_4 + 30r_2 r_3 r_4 + 24r_3^2 r_4 + 24r_1 r_4^2 \\
&\quad + 15r_2 r_4^2 + 24r_3 r_4^2 + 8r_4^3, \\
\mathrm{Vol}(S) &= (r_1 + r_3 + r_4)(5(r_1 + r_3 + r_4) + 4r_2).
\end{aligned}$$
(5.38)

Clearly, there is no decoupling limit. This can also be seen from the fact that S is not a rigid divisor. B is a $\mathbb{P}^1$ fibration over a toric dP_1 and S the reduction of this fibration over a non-rigid curve in this dP_1.

Computing the matter genera and the Yukawa numbers one finds for $SU(5)$:

$$g_{SU(6)} = 74, \quad g_{SO(10)} = 2, \quad n_{E_6} = 8, \quad n_{SO(12)} = 11. \tag{5.39}$$

and for $SO(10)$:

$$g_{SO(12)} = 9, \quad g_{E_6} = 5, \quad n_{E_7} = 16, \quad n_{SO(14)} = 52. \tag{5.40}$$

The fourfold X_4 is described by a reflexive polyhedron with 17 nef partitions, four of which describe an elliptic fibration over B. The Hodge numbers are collected in the table below:

type	$h^{1,1}$	$h^{2,1}$	$h^{3,1}$	χ
Tate	5	9	404	2448
$SU(5)$	13	9	84	360
$SO(10)$	17	11	43	342
$SU(5)_{U(1)}$	14	9	44	342
$SO(10)_{U(1)}$	18	11	39	324

(5.41)

Again, the Hodge numbers for $SU(5)/SO(10)$, without $U(1)$-restriction, do not fit the numbers calculated with formula (5.29). Examining the Tate equation after imposing the GUT group, we find an additional gauge enhancement at the divisor $y_4 = 0$. For $SU(5)$, the extra enhancement is also $SU(5)$, for $SO(10)$, the $y_4 = 0$ also carries an $SO(10)$ enhancement. Note that the second del Pezzo divisor in B, $y_2 = 0$, which is a dP_5, has a mathematical and a physical decoupling limit. It is a rigid divisor and the Euler numbers, after imposing the GUT groups on it, match the Euler numbers computed with (5.29). The form of the Tate equation implies that in that case no other divisor gets a non-abelian enhancement.

Finally, we consider an example of a dP_8 with a mathematical decoupling limit. The base geometry is given by the following weight matrix:

$$\begin{array}{c|ccccccc|c|c} & y_1 & y_2 & y_3 & y_4 & y_5 & y_6 & y_7 & \sum & \deg \\ \hline w_1 & 1 & 1 & 0 & 0 & 0 & 0 & 0 & 2 & 2 \\ w_2 & 1 & 0 & 1 & 1 & 1 & 0 & 0 & 4 & 3 \\ w_3 & 2 & 0 & 1 & 1 & 0 & 1 & 1 & 6 & 5 \end{array} \quad (5.42)$$

In the database, this model is labeled by `(cy4)p8v7n073d2-3-5t1`. There are two del Pezzo divisors: $y_2 = 0$ is a dP_0 and $y_5 = 0$, which we will name S, is dP_8. The existence of a mathematical decoupling limits can be deduced from the volumes of the base B and S:

$$\begin{aligned} \text{Vol}(B) &= 2r_1^3 + 15r_1^2 r_2 + 24r_1 r_2^2 + 11r_2^3 + 15r_1^2 r_3 + 60r_1 r_2 r_3 + 48r_2^2 r_3 + 30r_1 r_3^2 \\ &\quad + 60r_2 r_3^2 + 20r_3^3, \\ \text{Vol}(S) &= 4r_1 r_2 + 5r_2^2 + 8r_2 r_3. \end{aligned} \quad (5.43)$$

The mathematical decoupling limit can be implemented by setting $r_2 \to 0$. In that case, none of the other divisors will shrink to zero size. The topological data of the matter curves and the Yukawa couplings for $SU(5)$ models is

$$g_{SU(6)} = 38, \quad g_{SO(10)} = 0, \quad n_{E_6} = 2, \quad n_{SO(12)} = 4, \quad (5.44)$$

and for $SO(10)$:

$$g_{SO(12)} = 5, \quad g_{E_6} = 2, \quad n_{E_7} = 8, \quad n_{SO(14)} = 32. \quad (5.45)$$

Two nef-partitions are compatible with the elliptic fibration. The Hodge numbers and the Euler number are collected in the following table:

type	$h^{1,1}$	$h^{2,1}$	$h^{3,1}$	χ
Tate	4	26	182	1008
$SU(5)$	8	26	83	438
$SO(10)$	9	26	81	432
$SU(5)_{U(1)}$	9	26	71	372
$SO(10)_{U(1)}$	10	26	69	366

(5.46)

Even though there are no further non-abelian enhancements on the torically induced divisors of B, the Euler numbers do not match those obtained from (5.29). The mismatch might still be due to an extra non-abelian enhancement on a divisor that is not toric. Another possible explanation could be that we have a non-abelian enhancement over a curve. Resolving the singularities on these curves leads to a further Kähler parameter. However, we do not observe the corresponding Kähler modulus in the above table.

Three dP_7s

As a second class of examples, we discuss a model which has two different dP_7 divisors. The base is specified by the following weight matrix and hypersurface degrees:

	y_1	y_2	y_3	y_4	y_5	y_6	y_7	y_8	$\sum$	deg
w_1	1	1	0	0	0	0	0	0	2	2
w_2	1	0	1	1	0	1	0	1	5	4
w_3	1	0	0	0	1	1	0	0	3	2
w_4	0	0	1	1	0	0	1	0	3	2

(5.47)

The identifier for this model is (cy4)p9v8n152d2-4-2-2t2. The two dP_7s are given by $y_5 = 0$ and $y_7 = 0$, and we call the associated GUT branes S_5 and S_7. Let us first discuss the decoupling limits.

$$\begin{aligned}
\text{Vol}(B) &= 6r_1^2 r_2 + 6r_1 r_2^2 + 2r_2^3 + 6r_1^2 r_3 + 24r_1 r_2 r_3 + 12r_2^2 r_3 + 6r_1 r_3^2 + 6r_2 r_3^2 + 6r_1^2 r_4 \\
&\quad + 24r_1 r_2 r_4 + 12r_2^2 r_4 + 24r_1 r_3 r_4 + 24r_2 r_3 r_4 + 6r_3^2 r_4 + 12r_1 r_4^2 + 12r_2 r_4^2 \\
&\quad + 12r_3 r_4^2 + 4r_4^3 \,, \\
\text{Vol}(S_5) &= 2r_1^2 + 4r_1 r_3 + 4r_1 r_4 \,, \\
\text{Vol}(S_7) &= 4r_1 r_3 + 4r_2 r_3 + 2r_3^2 + 4r_3 r_4 \,.
\end{aligned}$$
(5.48)

As can be easily verified, S_5 has a mathematical as well as a physical decoupling limit, whereas S_7 only has a mathematical decoupling limit. The Kähler parameters can always be chosen in such a way that the respective GUT divisor is the only one whose volume goes to zero/remains finite in the mathematical/physical decoupling limit. The matter genera and Yukawa numbers for S_5 are the following:

$$g_{SU(6)} = 21\,, \quad g_{SO(10)} = 1\,, \quad n_{E_6} = 0\,, \quad n_{SO(12)} = 0\,, \tag{5.49}$$

for $SU(5)$, and

$$g_{SO(12)} = 3\,, \quad g_{E_6} = 1\,, \quad n_{E_7} = 4\,, \quad n_{SO(14)} = 24\,, \tag{5.50}$$

for $SO(10)$. As in the first dP_8-example, S_5 is not suitable for $SU(5)$ GUTs due to the absence of Yukawa points. For the divisor S_7 the topological data for $SU(5)$ and $SO(10)$ GUTs are as follows:

$$g_{SU(6)} = 48\,, \quad g_{SO(10)} = 0\,, \quad n_{E_6} = 2\,, \quad n_{SO(12)} = 4\,, \tag{5.51}$$

for $SU(5)$, and

$$g_{SO(12)} = 6\,, \quad g_{E_6} = 2\,, \quad n_{E_7} = 10\,, \quad n_{SO(14)} = 44\,, \tag{5.52}$$

for $SO(10)$. The associated Calabi-Yau fourfold has 25 nef-partitions, three of which describe an elliptic fibration over B. Imposing the GUT groups on S_5 (first block) and S_7 (second block), we compute the following Hodge numbers:

$$
\begin{array}{|c|cccc|}
\hline
\text{type} & h^{1,1} & h^{2,1} & h^{3,1} & \chi \\
\hline
\text{Tate} & 5 & 11 & 1066 & 1008 \\
SU(5) & 9 & 10 & 121 & 768 \\
SO(10) & 10 & 10 & 120 & 768 \\
SU(5)_{U(1)} & 10 & 10 & 78 & 516 \\
SO(10)_{U(1)} & 11 & 10 & 77 & 516 \\
\hline
SU(5) & 9 & 11 & 67 & 438 \\
SO(10) & 10 & 11 & 65 & 432 \\
SU(5)_{U(1)} & 10 & 11 & 55 & 372 \\
SO(10)_{U(1)} & 11 & 11 & 53 & 366 \\
\hline
\end{array}
\tag{5.53}
$$

For the $SU(5)$ and $SU(10)$ model on S_5, the Euler numbers agree with the formula (5.29) of [81], and there are also no further non-abelian enhancements in the Tate models. For S_7, there is a mismatch of Euler numbers, even though we do not find any further non-abelian enhancements on the toric divisors of B in the Tate model. However, there may be some singularities over non-toric divisors.

Now we would like to discuss a dP_7-model with a physical decoupling limit. For this purpose, we look at a base geometry which is specified by the following data:

$$
\begin{array}{|c|cccccccc|c|c|}
\hline
 & y_1 & y_2 & y_3 & y_4 & y_5 & y_6 & y_7 & y_8 & \sum & \deg \\
\hline
w_1 & 1 & 1 & 0 & 0 & 0 & 0 & 0 & 0 & 2 & 2 \\
w_2 & 1 & 0 & 1 & 1 & 0 & 0 & 0 & 0 & 3 & 2 \\
w_3 & 3 & 0 & 0 & 1 & 0 & 1 & 1 & 1 & 7 & 6 \\
w_4 & 2 & 0 & 0 & 0 & 1 & 1 & 1 & 0 & 5 & 4 \\
\hline
\end{array}
\tag{5.54}
$$

In the database, the label of this model is (cy4)p9v8n341d2-2-6-4t1. This model also has two dP_7's given by $y_3 = 0$ and $y_4 = 0$. The former has the same matter genera and Yukawa points as S_5 above, so we will focus on the latter which we will call S. The divisor S is not rigid. To see this, we have to examine B in more detail. We find that B is a dP_7 fibration over $\mathbb{P}^1$. Furthermore, the typical fiber of this fibration is equivalent to S. We note further that the divisor D_2 of the ambient space does not intersect the hypersurface, cf. section 5.1.2. The existence of a physical decoupling limit is inferred from the volumes of B and S:

$$
\begin{aligned}
\text{Vol}(B) &= 6r_1r_2^2 + 2r_2^3 + 24r_1r_2r_3 + 18r_2^2r_3 + 24r_1r_3^2 + 48r_2r_3^2 + 24r_3^3 + 24r_1r_2r_4 + 18r_2^2r_4 \\
&\quad + 48r_1r_3r_4 + 96r_2r_3r_4 + 120r_3^2r_4 + 24r_1r_4^2 + 48r_2r_4^2 + 120r_3r_4^2 + 40r_4^3, \\
\text{Vol}(S) &= 2(r_2 + 2r_3 + 2r_4)^2.
\end{aligned}
\tag{5.55}
$$

The physical decoupling limit is achieved when we take $r_1 \to \infty$, which is the volume of the $\mathbb{P}^1$, the base space of the fibration. In this limit, also the other dP_7 $y_3 = 0$, which is also a fiber, remains of finite size. The matter and Yukawa data for $SU(5)$ and $SO(10)$ GUTs are

$$
g_{SU(6)} = 57, \quad g_{SO(10)} = 1, \quad n_{E_6} = 4, \quad n_{SO(12)} = 6,
\tag{5.56}
$$

for $SU(5)$, and
$$g_{SO(12)} = 7, \quad g_{E_6} = 3, \quad n_{E_7} = 12, \quad n_{SO(14)} = 48, \tag{5.57}$$
for $SO(10)$. Extending the weight matrix of the base manifold we get an elliptically fibered Calabi-Yau fourfold which has 7 nef-partitions. Three of these are elliptic fibrations over B as given by (5.54). Computing the Hodge data, we get the following results:

type	$h^{1,1}$	$h^{2,1}$	$h^{3,1}$	χ
Tate	4	22	178	1008
$SU(5)$	12	22	53	306
$SO(10)$	16	24	50	300
$SU(5)_{U(1)}$	13	22	51	300
$SO(10)_{U(1)}$	17	24	48	294

(5.58)

Again, the Euler number from the Hodge data disagree with the one calculated from formula (5.29). Looking at the Tate model for the F-theory GUT, we find an additional $SU(5)$ or $SO(10)$-enhancement on the divisor $y_5 = 0$.

The toric three-/fourfold of [106]

The last example that we consider is the model (cy4)p8v7n080d1-1-3t1, which is equivalent to the compactification geometry discussed in [106], cf. also [107]. The base geometry is given by the following weight matrix and hypersurface:

	y_1	y_2	y_3	y_4	y_5	y_6	y_7	$\sum$	deg
w_1	1	1	0	0	0	0	0	2	1
w_2	0	1	1	1	0	0	0	3	1
w_3	0	2	1	0	1	1	1	6	3

(5.59)

This is an example of a base manifold that does not satisfy the almost Fano condition. The relevant dP_2 on which we will place the GUT model is D_4. Together with the dP_1 on D_1, these are the only two shrinkable del Pezzo surfaces as one can see from the volumes of B, $S = D_4$, and D_1,

$$\begin{aligned} \text{Vol} &= 3r_1^2 r_2 + 3r_1 r_2^2 + r_2^3 + 3r_1^2 r_3 + 18 r_1 r_2 r_3 + 9 r_2^2 r_3 + 12 r_1 r_3^2 + 18 r_2 r_3^2 + 10 r_3^3, \\ S &= (r_1 + 2r_3)^2, \\ D_1 &= r_2(2\, r_1 + r_2). \end{aligned} \tag{5.60}$$

Besides these two rigid del Pezzos, there are other toric dP_2's on $D_5 \sim D_6 \sim D_7$, which do not have a decoupling limit. Before we come to the fourfold geometry, we compute the matter and Yukawa data for $SU(5)$ and $SO(10)$ GUTs on S:

$$g_{SU(6)} = 134, \quad g_{SO(10)} = 0, \quad n_{E_6} = 6, \quad n_{SO(12)} = 10, \tag{5.61}$$

for $SU(5)$, and

$$g_{SO(12)} = 15, \quad g_{E_6} = 4, \quad n_{E_7} = 28, \quad n_{SO(14)} = 128, \tag{5.62}$$

for $SO(10)$.

Again, we extend the weight matrix of the base manifold to obtain an elliptically fibered Calabi-Yau fourfold which has 4 nef-partitions. Two of these are elliptic fibrations over B as given by (5.59). Computing the Hodge data for this fourfold and the reduced ones, we obtain the following results:

type	$h^{1,1}$	$h^{2,1}$	$h^{3,1}$	χ
Tate	4	0	2316	13968
$SU(5)$	8	0	1867	11298
$SO(10)$	9	0	1863	11280
$SU(5)_{U(1)}$	9	0	796	4878
$SO(10)_{U(1)}$	10	0	792	4860

(5.63)

These results match with the outcome of $SU(5)/SO(10)$ one finds from (5.29).

5.3 Summary and outlook

In this chapter, we have constructed a large class of Calabi-Yau fourfolds that are particularly useful for F-theory compactifications. There are several interesting directions for continued research.

Having such a large class of examples it might be useful to extend the rather basic analysis and to do more detailed calculations in F-theory. One possibility would be to include calculations with fluxes. It has been argued in [81, 87–89] that the spectral cover construction, which can be used to describe fluxes locally near the GUT brane [108], is valid in certain cases also beyond the local picture. Our data contains all the input needed to calculate chiral indices and tadpole cancellation conditions for a large class of models. Also the flux quantization and anomaly cancellation conditions worked out in [92, 93] could be included into the analysis.

In [100], F-theory models where the GUT brane does not wrap a del Pezzo divisor have been discussed. Despite the fact that the connection to many of the local GUT models discussed in the literature is not immediate, these GUTs are interesting because they may allow for gauge group breaking by discrete Wilson lines. The analysis we have performed for del Pezzo divisors can be extended to toric divisors in the base which are not del Pezzo.

So far no examples have been discussed where it is possible to make contact between F-theory GUT models and the Calabi-Yau fourfolds which are encountered in the calculation of $\mathcal{N} = 1$ superpotentials [109–114]. One could search our database for fourfold geometries which are suited for establishing a connection between these two exciting topics.

In our calculations we have made use of the software package PALP [1, 2], which can compute triangulations of polytopes and calculates the Mori cone, the Stanley-Reisner ideal and intersection rings for hypersurfaces in toric ambient spaces. An extension of these routines to the case of complete intersection Calabi-Yaus is interesting not only for applications in F-theory GUTs. Furthermore, it would be useful to extend PALP to handle also non-reflexive polytopes. In this context, the program cohomCalg [115] may be helpful for the calculation of

Hodge numbers. Finally, we should also try to overcome the problems with numerical overflows that arose due to the complexity of the fourfold polytopes.

A more mathematical question concerns methods to partially classify Calabi-Yau fourfolds. A complete classification of Calabi-Yau fourfolds that are hypersurfaces or complete intersections in a toric ambient space seems to be out of reach. An empirical formula due to H. Skarke [31] estimates the number N_d of reflexive polytopes in d dimensions to be of order $N_d \simeq 2^{2^{d+1}-4}$. This implies that the number of reflexive polytopes in 5 dimension is of order $\mathcal{O}(10^{18})$. In 6 dimensions there are even expected to be $\mathcal{O}(10^{37})$ reflexive polytopes. Since also non-reflexive polytopes may be of interest in F-theory, this number might only be the tip of the iceberg. Even a classification of elliptically fibered Calabi-Yau fourfolds may be too difficult. However, what could in principle be doable is a complete classification of the geometries we have constructed in this chapter. The prescription is the following: take each of the 473 800 776 reflexive polyhedra in four dimensions and put in all possible hypersurfaces whose degree is below the degree of the Calabi-Yau hypersurface in this ambient space. Then construct fourfolds that are elliptic fibrations over these base manifolds. A naïve estimate shows that this procedure would yield $\mathcal{O}(10^{11})$ fourfold geometries. Due to the overflow problem we can only claim that we have a full classification of this type of Calabi-Yau fourfolds if they originate from reflexive polyhedra in four dimensions which have up to seven points.

Chapter 6

Restrictions on infinite sequences of type IIB vacua

With our present understanding, string theory seems to allow for a vast number of metastable four-dimensional vacua. This set of universes is often called the string landscape [116], and is equipped with a, in principle computable, effective potential. In a scenario where our universe is described by fluctuations around a particular minimum of this potential, particle masses and couplings are given by local curvatures at the minimum. But there might also be more subtle observational effects depending on the large scale structure of the potential.

One important topographical feature, relevant for many effects in string cosmology, is the existence of sequences of vacua connected by continuous potential barriers. When quantum effects are taken into account, tunnelling can occur between the vacua, with a probability that is computable once the features of the potential barrier are known. From the space-time perspective, the tunnelling process consists of the nucleation of a bubble of the new vacuum inside the old vacuum phase. Depending on the tunnelling rate, and the expansion rates of the new and old universes, the transition to the new vacuum is either complete or partial. The latter case, when part of space-time remains in the old vacuum, is known as eternal inflation [117–119]. Potentially, bubble collisions in such a cosmological scenario could leave an observable imprint in the CMBR [120, 121], and this was recently compared with the WMAP 7-year results [122, 123].

Another interesting possibility is that of chain inflation [124–127], see also [128]. In this kind of models, inflation results from sequential tunnelling in a chain of de Sitter universes, each supporting just part of an e-folding. To be a viable option, chain inflation requires the existence of sequences of neighboring vacua with certain properties. Other effects hinging on our local landscape surroundings are resonance tunnelling [129], "giant leaps" between far-away vacua [130], and disappearing instantons [131, 132] – effects that can greatly affect tunnelling probabilities. In all these cases, detailed knowledge of the potential is required to obtain quantitative results.

One part of the landscape that offers fairly accurate analytical and numerical control is the complex structure moduli space of type IIB flux compactifications. Fluxes piercing non-trivial three-cycles of the internal geometry generate a potential with discrete minima for the complex structure moduli and axio-dilaton. Each of these minima corresponds, after fixing of the Kähler moduli [3, 47, 133], to a vacuum in the landscape. The reason for the mathematical tractability of many of these models is that the internal manifold remains conformally Calabi-Yau after the introduction of fluxes, making the powerful tool-kit of special geometry applicable. Indeed, as demonstrated in Ref. [11, 134], the potential of the resulting four-dimensional $\mathcal{N} = 1$ supergravity is determined by the Gukov-Vafa-Witten superpotential [51] making it straightforward to compute.

Taking advantage of this fact, type IIB flux compactifications and the resulting potential have been studied in many contexts. In Ref. [135] it was shown that D3-brane black holes, which also affect the potential for the moduli, must be small to affect the minimization, but can then potentially serve as seeds for bubble nucleation, and in Ref. [136], explicit profiles of BPS domain walls interpolating between different vacua were obtained. Exploiting the fact that fluxes transform under monodromy transformations, it was demonstrated in [137, 138] that long sequences of continuously connected vacua are a common feature in the landscape, thus opening up the possibility of chain inflation or resonance tunnelling in this framework. These studies were extended in [139], where the tunnelling probabilities between vacua in the sequence were first computed. Moreover, Ref. [131, 140] investigated the influence of the universal Kähler modulus on tunnelling rates and domain wall dynamics in this setting.

Recently, Ahlqvist et al. [141] continued these investigations of type IIB vacuum sequences by a thorough study of a class of one-parameter Calabi-Yau models. This revealed several intriguing features, both in the tunnelling dynamics and in the vacuum structure. For vacua connected by conifold monodromies, it was demonstrated that the tunnelling trajectories tend to pass close to the conifold point. Furthermore, long sequences of connected minima seemingly accumulating to the large complex structure (LCS) point were found. It was left as an open question whether these sequences continue indefinitely or not. That the sequences approach the LCS point is particularly interesting in view of the no-go theorem [5] derived by Ashok and Douglas stating that infinite sequences of vacua with imaginary self-dual (ISD) flux can only occur if they accumulate to so-called D-limits – one example of which is the LCS point. It is the aim of the present chapter to investigate if it is possible to have infinite sequences of minima accumulating to the LCS point, and in particular to determine whether the sequences found in [141] end or not.

Previous studies [142–145] of the finiteness of type IIB flux vacua have mainly been based on the statistical methods pioneered in [146, 147]. (For reviews, consult e.g. [7, 12].) This approach uses a continuum approximation of the fluxes, which allows to relate the density of vacua in moduli space to the Euler density of a certain metric on moduli space. As this Euler density is an index, it only gives a lower bound for the true vacuum density in principle. The index can be shown to integrate to finite values around regular points in moduli space [5]. Its structure around Calabi-Yau singularities has been analyzed in [144, 145], including both ADE singularities and the LCS points. In both cases the index integrates to finite results. For results concerning the finiteness of the intersecting D-brane landscape, see e.g. [148–150].

In this work we complement these statistical studies with a more direct analysis of the possibility of having any infinite sequence of ISD vacua. By analyzing the geometry of the complex structure moduli space we flesh out the details of the Ashok-Douglas theorem and obtain explicit expressions for a positive definite quadratic form that must stay finite in any sequence of ISD vacua. Using this, we derive an extension of the no-go theorem. Namely, for one-parameter Calabi-Yau manifolds, there are no sequences of vacua accumulating to the LCS point. This shows by analytical means that, in particular, the sequences of [141] are finite. We furthermore extend this result to the D-limits corresponding to conifold points and decoupling limits, and also study the LCS limit of a two-parameter model.

In addition, we treat the case when the compactification manifold is $K3 \times K3$, for which the finiteness is proven in a very different manner [151]. We find that D-limits exist and that infinite sequences can be constructed. Hence all but a finite number of the solutions in a sequence must be related by automorphisms of $K3$. We demonstrate this in a simple example. Finally, we use numerical methods to study two particular examples of one-parameter Calabi-Yau manifolds. Using expansions of the periods in the LCS region allows us to efficiently compute the scalar potential, and thus follow the sequences of [141] closer to the LCS point. In accordance with the general analysis, the minima eventually leave the region close to the LCS point.

This chapter is organized as follows. Section 6.1 contains a review on type IIB compactifications and introduces our notation and conventions. We then discuss the no-go theorem by Ashok and Douglas and the relevance of D-limits in section 6.2. Subsequently, we analyze the length of sequences of ISD vacua in various D-limits in type IIB compactifications. Details of this computation is relegated to appendix C.1. In section 6.4, we discuss sequences of vacua in D-limits in F-theory compactifications on $K3 \times K3$. Through a numerical analysis, we then map out two sequences of type IIB vacua in the LCS region in section 6.5. Finally, we summarize and discuss our results.

6.1 Type IIB moduli stabilization

In this section we give a brief review of moduli stabilization in type IIB supergravity, and set the notation and conventions. To aid comparison with that work we use as far as possible the notation of [141].

6.1.1 Calabi-Yau geometry

We denote by $\mathcal{M}$ a Calabi-Yau manifold with complex structure moduli space M, and let $\mathcal{C}$ be the combined moduli space of complex structure and the axio-dilaton: $(z, \tau) \in \mathcal{C}$. The periods of $\mathcal{M}$ are

$$\Pi_I = \int_{C_I} \Omega = \int_{\mathcal{M}} C_I \wedge \Omega, \tag{6.1}$$

where Ω is the holomorphic three-form and C_I is a basis in $H_3(\mathcal{M})$. Note that C_I is used to denote two things: both the cycles and their Poincaré duals. The intersection matrix $Q = (Q_{IJ})$ is defined by

$$Q_{IJ} = \int_{C_I} C_J = \int_{\mathcal{M}} C_I \wedge C_J. \tag{6.2}$$

The periods are collected into a vector whose entries we number in reverse order

$$\Pi(z) = \begin{pmatrix} \Pi_N(z) \\ \Pi_{N-1}(z) \\ \vdots \\ \Pi_0(z) \end{pmatrix}, \tag{6.3}$$

where z is a $h^{(1,2)}$-dimensional (complex) coordinate on $\mathcal{M}$ and $N \equiv 2h^{1,2} + 1$.

In our one-parameter examples there are three special points in moduli space: the large complex structure (LCS) point, the conifold point and the Landau-Ginzburg point. We fix these to lie at $z = 0, 1$ and $z = \infty$, respectively. The periods are subject to monodromies upon transport around these points:

$$\Pi \to T \cdot \Pi, \tag{6.4}$$

where T is a matrix that preserves the symplectic structure Q. The complex structure moduli space is furthermore equipped with a Kähler metric, with Kähler potential

$$K_{\text{cs}} = -\log\left(i \int_{\mathcal{M}} \Omega \wedge \bar{\Omega}\right) = -\log(i \Pi^\dagger \cdot Q^{-1} \cdot \Pi). \tag{6.5}$$

Note that our integration conventions are such that

$$\int_{\mathcal{M}} \bar{\Gamma} \wedge *\Gamma > 0 \tag{6.6}$$

for any non-zero three-form Γ. Finally, we define [152] the (antisymmetric, topological, moduli independent) intersection product and the (symmetric, positive definite, moduli dependent) scalar product as

$$\langle A_{(3)}, B_{(3)} \rangle = \int_{\mathcal{M}} A_{(3)} \wedge B_{(3)} = A \cdot Q \cdot B^T, \tag{6.7}$$

$$(A_{(3)}, B_{(3)}) = \int_{\mathcal{M}} A_{(3)} \wedge *B_{(3)} = A \cdot \mathcal{G}_z \cdot B^T, \tag{6.8}$$

respectively. Here, $A = (A^0, \ldots A^N)$ is a row vector collecting the components of the form $A_{(3)}$ in the basis C_I: $A_{(3)} = -A^I C_I$, and similarly for B. The matrix $\mathcal{G}_z$ is a moduli dependent positive quadratic form on $\mathbb{C}^{N+1}$.

6.1.2 Flux vacua

Fluxes piercing the three-cycles induce a Gukov-Vafa-Witten superpotential W, leading to an $\mathcal{N} = 1$ scalar potential potentially stabilizing all complex structure moduli and the axio-dilaton.

The potential is

$$V(z,\tau) = e^K \left(g^{i\bar{j}} D_i W D_{\bar{j}} \bar{W} + g^{\tau\bar{\tau}} D_\tau W D_{\bar{\tau}} \bar{W} + g^{\rho\bar{\rho}} D_\rho W D_{\bar{\rho}} \bar{W} - 3|W|^2 \right), \qquad (6.9)$$

where $g^{A\bar{B}} = (\partial_A \partial_{\bar{B}} K)^{-1}$ and $D_A = \partial_A + \partial_A K$ with K being the $\mathcal{N}=1$ Kähler potential. To compute V all that is needed are expressions for the superpotential W and the Kähler potential K. We denote the three-form fluxes by $F_{(3)}$ (R-R) and $H_{(3)}$ (NS-NS). We collect the flux quanta in row vectors $F = (F^0, \ldots, F^N)$ defined by

$$F^I = -(Q^{-1})^{IJ} \int_{C_J} F_{(3)}, \qquad (6.10)$$

and similarly for H. Note that the vectors F and H are subject to Dirac quantization. Their entries are integer multiples of $4\pi^2 \alpha'$ which we fix to unity for convenience. For notational convenience we often use the combined three-form flux

$$G_{(3)} = F_{(3)} - \tau H_{(3)}, \qquad (6.11)$$

which can also be represented by a vector, albeit with non-integer components

$$G = F - \tau H. \qquad (6.12)$$

The superpotential is given by

$$W = \int_{\mathcal{M}} \Omega \wedge G_{(3)} = G \cdot \Pi. \qquad (6.13)$$

The Kähler potential is

$$K = -\ln\left(-i(\tau - \bar{\tau})\right) + K_{\text{cs}}(z, \bar{z}) - 3\ln\left(-i(\rho - \bar{\rho})\right), \qquad (6.14)$$

where K_{cs} is the Kähler potential on complex structure moduli space, given by (6.5). Due to the last term in the Kähler potential, the contributions of $g^{\rho\bar{\rho}} D_\rho W D_{\bar{\rho}} \bar{W}$ and $-3|W|^2$ cancel:

$$V(z,\tau) = e^K \left(g^{i\bar{j}} D_i W D_{\bar{j}} \bar{W} + g^{\tau\bar{\tau}} D_\tau W D_{\bar{\tau}} \bar{W} \right). \qquad (6.15)$$

Using (6.13), (6.14) and (6.5), the scalar potential can be computed numerically once the periods and their derivatives are known.

The three-form fluxes induce a D3-brane charge density, that must be compensated by localized sources on the compact manifold. This amounts to the tadpole condition

$$\int_{\mathcal{M}} F_{(3)} \wedge H_{(3)} = \frac{\chi}{24} + \tfrac{1}{4} N_{\text{O}3} - N_{\text{D}3} \equiv L, \qquad (6.16)$$

where $N_{\text{O}3}$ is the number of O3 planes, $N_{\text{D}3}$ is the number of (space filling) D3 branes in the compactification and χ counts the tadpole contribution of D7 branes and O7-planes. From the F-theory perspective, χ is the Euler characteristic of (an appropriate resolution of) the corresponding elliptic Calabi-Yau four-fold. Expressed in the vectors F and H the tadpole condition reads

$$F \cdot Q \cdot H^T = L. \qquad (6.17)$$

Giddings, Kachru and Polchinski [11] showed how to solve all equations of motion in the above set-up (see also [134]). For sources satisfying a certain "BPS-like" condition and at tree level, the equations forces the flux to be imaginary self dual (ISD): $*G_{(3)} = iG_{(3)}$. This is a condition on the complex structure moduli and the axio-dilaton. In fact, the ISD condition is equivalent to the vanishing of the F-terms related to these moduli. We have the equivalences

$$\begin{aligned} \text{EOMs} &\iff *G_{(3)} = iG_{(3)} \iff G_{(3)} \in H^{(2,1)}(\mathcal{M}) \oplus H^{(0,3)}(\mathcal{M}) , \\ D_\tau W &= 0 \iff G_{(3)}^{(3,0)} = 0 , \\ D_i W &= 0 \iff G_{(3)}^{(1,2)} = 0 , \\ D_\rho W &= 0 \iff W = 0 \iff G_{(3)}^{(0,3)} = 0 , \end{aligned} \qquad (6.18)$$

where the harmonic representative is understood by the $\in$ in the first line. We see that ISD implies that $D_\tau W = D_i W = 0$ but that supersymmetry can well be broken by a non-zero $D_\rho W$ in ISD minima.

Note furthermore that if z and τ are tuned so that the flux is ISD, then the potential (6.15) has a global minimum. These minima are discrete, and each such configuration corresponds to an ISD vacuum of the type IIB landscape.

6.2 Series in D-limits

Let us now turn to the question of the existence of infinite sequences of ISD vacua. Ashok and Douglas [5] have formulated a no-go theorem that restricts the possibility of infinite sequences of vacua. They also demonstrated that this theorem can be evaded in the vicinity of special points – so-called D-limits – in the moduli space $\mathcal{C}$, one example being the point of large complex structure. In this section we review and make this theorem and the concept of D-limits more precise.

6.2.1 The no-go theorem of Ashok and Douglas

The two main ingredients of the argument are the tadpole and the ISD conditions. Suppose that we have no anti-D3-branes, and that the flux $G_{(3)}$ is imaginary self dual. These conditions include all supersymmetric vacua, but as explained above also other minima. We shall keep these two assumptions throughout this section. We then have that

$$\langle F_{(3)}, H_{(3)} \rangle = \frac{\chi}{24} + \tfrac{1}{4} N_{O3} - N_{D3} = L , \qquad (6.19)$$

with L being a number bounded from above since $N_{D3} \geq 0$. On the other hand a short computation yields

$$\langle F_{(3)}, H_{(3)} \rangle = \frac{i}{2 \operatorname{Im} \tau} \langle \bar{G}_{(3)}, G_{(3)} \rangle = \frac{1}{2 \operatorname{Im} \tau} \langle \bar{G}_{(3)}, *G_{(3)} \rangle = \frac{1}{2 \operatorname{Im} \tau} \bar{G} \cdot \mathcal{G}_z \cdot G^T > 0 , \qquad (6.20)$$

where, in the second step, the imaginary self-duality of $G_{(3)}$ was used. So, in fact, for our type of vacua we have

$$\frac{1}{2 \operatorname{Im} \tau} \bar{G} \cdot \mathcal{G}_z \cdot G^T = L \leq L_{\max} = \frac{\chi}{24} + \tfrac{1}{4} N_{O3} . \qquad (6.21)$$

Since $\mathcal{G}_z$ is a positive quadratic form we have thus shown that G must lie inside a moduli dependent ellipsoid in $\mathbb{C}^{N+1}$. Let us use this to derive a restriction on the integer valued vectors F and H. If we collect these into a $(2N+2)$-dimensional vector $\hat{N} = (F, H)$ we have

$$\langle F_{(3)}, H_{(3)} \rangle = \frac{1}{2\operatorname{Im}\tau} \bar{G} \cdot \mathcal{G}_z \cdot G^T = \hat{N} \cdot (\mathcal{G}_\tau \otimes \mathcal{G}_z) \cdot \hat{N}^T , \quad (6.22)$$

where $\mathcal{G}_\tau$ is proportional to the metric on a torus with complex structure τ:

$$\mathcal{G}_\tau = \frac{1}{2\operatorname{Im}\tau} \begin{pmatrix} 1 & -\operatorname{Re}\tau \\ -\operatorname{Re}\tau & |\tau|^2 \end{pmatrix}. \quad (6.23)$$

Thus, the integer vector $\hat{N}$ must lie within an ellipsoid in $\mathbb{R}^{2N+2}$ whose dimensions are given by the (z, τ)-dependent eigenvalues Λ_i of the matrix $\mathcal{G}_\tau \otimes \mathcal{G}_z$. It is now simple to formulate the no-go result of [5]. Any region $R \subset \mathcal{C}$ of (τ, z)-space for which the Λ_i are bounded from below by some positive number, can support only a finite number of vacua. To see this, suppose that $\Lambda_i(z, \tau) > \epsilon$ for all (z, τ) and i. Then all admissible $\hat{N}$ lie within a ball of radius squared $r^2 = L_{\max}/\epsilon$. These are of course finitely many.

It is also immediately clear how to evade this no-go result. Infinite series of vacua can occur only if their location in $\mathcal{C}$ approaches a point where the matrix $\mathcal{G}_\tau \otimes \mathcal{G}_z$ develops a null eigenvector. Points where this happens are referred to as D-limits.

6.2.2 D-limits

Since the eigenvalues of a product matrix is the product of the eigenvalues of the factors, a D-limit can arise in two ways. Either $\mathcal{G}_\tau$ or $\mathcal{G}_z$ can degenerate. In the first case, using S-duality to restrict τ to lie in the standard fundamental domain of the torus moduli space, the only locus where $\mathcal{G}_\tau$ degenerates is as $\operatorname{Im}\tau \to \infty$. This limit corresponds to a decoupling limit, and the null eigenvector has only R-R flux.

The other option is that $\mathcal{G}_z$ degenerates. To find out when this happens we need to compute this matrix in terms of the periods. Using the expression for $(A_{(3)}, B_{(3)})$ given in Eq. (2.18) of [152] some simple algebra yields

$$A \cdot \mathcal{G}_z \cdot B^T = 2e^K \operatorname{Re} \left[(A \cdot \Pi)(\Pi^\dagger \cdot B^T) + g^{i\bar{j}}(A \cdot D_i \Pi)(\bar{D}_j \Pi^\dagger \cdot B^T) \right] , \quad (6.24)$$

so that

$$\mathcal{G}_z = 2e^K \operatorname{Re} \left[\Pi \Pi^\dagger + g^{i\bar{j}} D_i \Pi \bar{D}_j \Pi^\dagger \right] . \quad (6.25)$$

This matrix can be computed straightforwardly when the periods Π_I are known. In section 6.3 we shall do this for the large complex structure and conifold limits.

6.2.3 D-limits and F-theory

Flux compactifications of Type IIB string theory can be embedded in the more general framework of F-theory compactified on elliptic Calabi-Yau fourfolds, see e.g. [12]. On the one hand,

F-theory geometrizes the $SL(2,\mathbb{Z})$ self-duality of type IIB string theory. For generic points in moduli space, F-theory models have no interpretation in terms of perturbative type IIB string theory due to the presence of various types of (p,q)-branes. In Sen's weak coupling limit [14], however, F-theory reduces to weakly coupled type IIB string theory compactified on Calabi-Yau orientifolds with O7-planes and D7-branes. In F-theory, the closed string moduli are unified with the open string moduli in the moduli space of the elliptic Calabi-Yau manifold. On the other hand, F-theory can be obtained as a limit of M-theory compactifications on elliptic Calabi-Yau manifolds by collapsing the elliptic fiber. As M-theory contains a four-form field strength, one can introduce four-form fluxes $G_{(4)}$. These must have one leg in the elliptic fiber in order not to spoil Lorentz invariance [153]. In Sen's weak coupling limit, the four-form fluxes $G_{(4)}$ on the M-theory side encode both the three-form flux $G_{(3)}$ as well as (abelian) two-form fluxes $F_{(2)}$ on D7-branes on the type IIB side.

The analysis of the last section can be carried over to this case: In the absence of $O3$-planes, the condition for the cancellation of the $D3$-brane tadpole is

$$\frac{\chi(CY_4)}{24} - \frac{1}{2}\int_{CY_4} G_{(4)} \wedge G_{(4)} = N_{D3}. \tag{6.26}$$

Here, $\chi(CY_4)$ denotes the Euler characteristic of the elliptic Calabi-Yau fourfold.

As shown in [154], the equations of motion enforce that

$$*G_{(4)} = G_{(4)}, \tag{6.27}$$

so that

$$\frac{1}{2}\int_{CY_4} G_{(4)} \wedge G_{(4)} = \frac{1}{2}\int_{CY_4} G_{(4)} \wedge *G_{(4)} \geq 0. \tag{6.28}$$

As before, infinite sequences of flux vacua can only exist in a limit in which this positive definite form develops a zero eigenvector.

6.3 Series in type IIB D-limits

In this section we analyze the possibility of infinite sequences in various D-limits in type IIB compactifications. We treat in turn the large complex structure limit, decoupling limits and the conifold limit. We assume all the time that only one of these special loci is approached, i.e., we do not treat a simultaneous decoupling and LCS limit. In all cases we find that no infinite sequences of ISD vacua are possible.

6.3.1 Series around a large complex structure point

An example of a D-limit that is ubiquitous in Calabi-Yau moduli spaces is the large complex structure point. Since the series of [141] seem to accumulate at this point it is natural to investigate whether such series can continue indefinitely or not. We study therefore one-parameter models with an LCS point and use the no-go results of Ashok and Douglas. The complex structure modulus is conventionally denoted by $t = t_1 + it_2$ with $t_{1,2} \in \mathbb{R}$ and the LCS point is at $t_2 \to \infty$. For a one-parameter model, the period vector takes the following general form around the LCS point

$$\begin{pmatrix} \Pi_3 \\ \Pi_2 \\ \Pi_1 \\ \Pi_0 \end{pmatrix} \sim \begin{pmatrix} \alpha_3 t^3 + \gamma_3 t + i\delta_3 \\ \beta_2 t^2 + \gamma_2 t + \delta_2 \\ t \\ 1 \end{pmatrix}. \tag{6.29}$$

Using Eqs. (6.25), (6.5), the definition $g_{i\bar{j}} = \partial_i \partial_{\bar{j}} K$, and the expansion of the periods now allows for a straightforward computation of $\mathcal{G}_t$. The computation is outlined in Appendix C.1, and a generic[1] model of our type gives the result

$$\mathcal{G}_t = \begin{pmatrix} a_{11} t_2^3 + \mathcal{O}(t_2) & a_{12} t_2 + \mathcal{O}(1/t_2) & a_{13} \frac{1}{t_2} + \mathcal{O}(1/t_2^3) & a_{14} \frac{1}{t_2^3} + \mathcal{O}(1/t_2^5) \\ \cdot & a_{22} t_2 + \mathcal{O}(1/t_2) & a_{23} \frac{1}{t_2} + \mathcal{O}(1/t_2^3) & a_{24} \frac{1}{t_2^3} + \mathcal{O}(1/t_2^5) \\ \cdot & \cdot & a_{33} \frac{1}{t_2} + \mathcal{O}(1/t_2^3) & a_{34} \frac{1}{t_2^3} + \mathcal{O}(1/t_2^5) \\ \cdot & \cdot & \cdot & a_{44} \frac{1}{t_2^3} + \mathcal{O}(1/t_2^5) \end{pmatrix}, \tag{6.30}$$

for some known constants a_{ij}. Here, the entries $\cdot$ in $\mathcal{G}_t$ are determined by symmetry. It is clear that this matrix develops two null eigenvectors as $t_2 \to \infty$.

We now prove that there are no infinite series of ISD vacua accumulating at the complex structure point for our one-parameter models. Let us begin by noting that the intersection matrix in the basis of (6.29) is anti-diagonal:

$$Q_{03} = -Q_{12} = -1. \tag{6.31}$$

Therefore a flux configuration with $F^0 = F^1 = H^0 = H^1 = 0$ satisfies

$$\langle F_{(3)}, H_{(3)} \rangle = 0, \tag{6.32}$$

implying that for any ISD vacuum corresponding to such fluxes

$$\hat{N} \cdot (\mathcal{G}_\tau \otimes \mathcal{G}_t) \cdot \hat{N}^T = 0. \tag{6.33}$$

Since the matrix $\mathcal{G}_t$ is positive definite for any smooth manifold, (6.33) implies that the compactification manifold is singular, i.e., that the vacuum sits exactly at the D-limit. In fact, as remarked in [137], the flux potential always has a minimum at the LCS point for such flux configurations. What we shall demonstrate below is that this is the only possibility: there are no series for which one of $F^{0,1}$, $H^{0,1}$ is nonzero.

Since we assume that $\operatorname{Im} \tau$ stays finite, the essential features can be deduced from the structure of $\mathcal{G}_t$. We prove first the following statement. Suppose $\{N_n = (N_n^0, N_n^1, N_n^2, N_n^3)\}$ is

---

[1] We assume that two of the expansion coefficients in (6.29) are related as $\beta_2 = 3\alpha_3$. This is true for all models in [141] and seems to be a general feature. Treating the case $\beta_2 \neq 3\alpha_3$ produces results identical to those presented here.

a series of integer four-vectors and that $\{t_n\}$ is a series of points in complex structure moduli space such that

$$\lim_{n\to\infty} \operatorname{Im} t_n = \infty, \qquad \lim_{n\to\infty} N_n \cdot \mathcal{G}_{t_n} \cdot N_n^T \equiv \lim_{n\to\infty} \|N_n\|_t^2 \neq \infty. \tag{6.34}$$

Then $N_n^0 = N_n^1 = 0$ for n sufficiently large. We prove this by contradiction. To reduce clutter, let us from now on suppress the subscript on t and N. Suppose first that N^0 is non-zero, without loss of generality let $N^0 = 1$ and assume that (6.34) holds. Denote the eigenvectors and (positive) eigenvalues of the matrix (6.30) by w_i and λ_i, respectively. The scalar product $\|N\|_t^2$ can be expanded in this eigenbasis:

$$\|N\|_t^2 = \sum_{i=1}^4 |N \cdot w_i(t)|^2 \lambda_i(t). \tag{6.35}$$

Since all of the terms in this expression are positive, all of them must stay finite in the limit $t_2 \to \infty$. Expanding in t_2, the eigenvalues and eigenvectors are given by

$$\begin{aligned}
\lambda_1 &= a_{11} t_2^3 + \mathcal{O}(t_2), & w_1^T &= [1, \mathcal{O}(t_2^{-2}), \mathcal{O}(t_2^{-4}), \mathcal{O}(t_2^{-6})], \\
\lambda_2 &= a_{22} t_2 + \mathcal{O}(t_2^{-1}), & w_2^T &= [\mathcal{O}(t_2^{-2}), 1, \mathcal{O}(t_2^{-2}), \mathcal{O}(t_2^{-4})], \\
\lambda_3 &= \frac{a_{33}}{t_2} + \mathcal{O}(t_2^{-2}), & w_3^T &= [\mathcal{O}(t_2^{-4}), \mathcal{O}(t_2^{-2}), 1, \mathcal{O}(t_2^{-2})], \\
\lambda_4 &= \frac{a_{44}}{t_2^3} + \mathcal{O}(t_2^{-5}), & w_4^T &= [\mathcal{O}(t_2^{-6}), \mathcal{O}(t_2^{-4}), \mathcal{O}(t_2^{-2}), 1].
\end{aligned} \tag{6.36}$$

The first eigenvalue grows as $\lambda_1 \sim t_2^3$. Therefore we must have $|N \cdot w_1|^2 \sim \mathcal{O}(t_2^{-3})$. Hence

$$\mathcal{O}\left(t_2^{-3/2}\right) = N \cdot w_1 = 1 + N^1 \mathcal{O}(t_2^{-2}) + N^2 \mathcal{O}(t_2^{-4}) + N^3 \mathcal{O}(t_2^{-6}). \tag{6.37}$$

This can happen only if at least one of the N^i diverges. It is also clear that this must happen in order for N to approach one of the zero eigenvectors of $\mathcal{G}_t$. What is needed is

$$N^1 = P t_2^2 + o(t_2^2), \qquad N^2 = Q t_2^4 + o(t_2^4), \qquad N^3 = R t_2^6 + o(t_2^6), \tag{6.38}$$

where, e.g., $o(t_2^2)$ denotes terms that grows slower than t_2^2 and P, Q and R are appropriately chosen constants. Consider now the term in (6.35) proportional to λ_4. We obtain

$$N \cdot w_4 = \mathcal{O}(t_2^{-6}) + N^1 \mathcal{O}(t_2^{-4}) + N^2 \mathcal{O}(t_2^{-2}) + N^3 \sim R t_2^6. \tag{6.39}$$

Hence $|N \cdot w_4|^2 \lambda_4 \sim R^2 t_2^9$, and R must vanish. Considering now in order the terms proportional to λ_3 and λ_2 demonstrates in complete parallel that also Q and P must be zero. This is, however, incompatible with (6.37), and we have reached a contradiction. We have thus proved that $N^0 = 0$. Assuming now a flux of the form $N = (0, 1, N^2, N^3)$ and going through an almost identical argument demonstrates $N^1 = 0$.

To complete the argument we now consider a series of integer eight-vector $\hat{N}_n = (F_n, H_n)$ and assume that

$$\hat{N} \cdot \mathcal{G}_\tau \otimes \mathcal{G}_t \cdot \hat{N}^T, \tag{6.40}$$

stays finite as the LCS point is approached. (Again we suppress the index on $\hat{N}$ and t.) We furthermore assume that $\tau = \tau_1 + i\tau_2$ lies in the standard fundamental domain $|\tau| \geq 1$, $|\tau_1| \leq 1/2$. Since we, by assumption, do not approach a decoupling limit, the whole series fulfills $\tau_2 \leq M$ for some positive number M. This means that the eigenvalues $\mu_{1,2}$ of the matrix (6.23) are bounded from below. The eigenvalues and orthonormal eigenvectors $v_{1,2}$ are

$$\mu_{1,2} = \frac{1}{4\tau_2}\left[1 + |\tau|^2 \pm \sqrt{(1-|\tau|^2)^2 + 4\tau_1^2}\right], \qquad v_{1,2} = \begin{pmatrix} v_{1,2}^F \\ v_{1,2}^H \end{pmatrix}. \qquad (6.41)$$

The only property of $v_{1,2}$ important to us presently is their orthonormality. To see that the eigenvalues are bounded note that

$$\mu_1 \geq \mu_2 \geq \frac{1}{4\tau_2}\left(2 - \sqrt{4\tau_1^2}\right) = \frac{1}{2\tau_2}(1-\tau_1) > \frac{1}{2M}(1-\tau_1) \geq \frac{1}{4M}. \qquad (6.42)$$

We can now expand in eigenvectors

$$\hat{N} \cdot \mathcal{G}_\tau \otimes \mathcal{G}_t \cdot \hat{N}^T = \sum_{i,j} |\hat{N} \cdot v_i \otimes w_j|^2 \mu_i \lambda_j = \sum_{i,j} |\epsilon_{ij}|^2 \lambda_j \mu_i, \qquad (6.43)$$

where

$$\epsilon_{ij} \equiv \hat{N} \cdot v_i \otimes w_j = v_i^F (F \cdot w_j) + v_i^H (H \cdot w_j). \qquad (6.44)$$

Again, each term in the sum (6.43) has to stay finite in the limit. Since the μ_i are bounded, the quantities ϵ_{1j} and ϵ_{2j} must each satisfy

$$\epsilon_{ij} = \mathcal{O}(1/\sqrt{\lambda_j}). \qquad (6.45)$$

Using the orthonormality of $v_{1,2}$ (6.44) is easily inverted. In matrix notation

$$\begin{pmatrix} F \cdot w_j \\ H \cdot w_j \end{pmatrix} = \begin{pmatrix} v_1^F & v_2^F \\ v_1^H & v_2^H \end{pmatrix} \begin{pmatrix} \epsilon_{1j} \\ \epsilon_{2j} \end{pmatrix}. \qquad (6.46)$$

Since the $v_i^{F/H}$ are bounded non-zero numbers we have therefore proven that

$$F \cdot w_j = \mathcal{O}(1/\sqrt{\lambda_j}) \qquad H \cdot w_j = \mathcal{O}(1/\sqrt{\lambda_j}). \qquad (6.47)$$

This is exactly what is needed to prove that $F^0 = F^1 = H^0 = H^1 = 0$ from the structure of $\mathcal{G}_t$, starting from Eq. (6.37).

To sum up, requiring the finiteness of $\hat{N} \cdot \mathcal{G}_\tau \otimes \mathcal{G}_t \cdot \hat{N}^T$ in the limit $t_2 \to \infty$ implies that $F^0 = F^1 = H^0 = H^1 = 0$. This in turn implies that there is no vacuum, except the singular one located exactly at the LCS point.

6.3.2 Series in decoupling limits

Let us now, in a very similar manner, prove that there can be no sequences of ISD vacua converging to a decoupling limit. We consider some flux compactification on a Calabi-Yau whose matrix $\mathcal{G}_z$ has eigenvalues and vectors λ_j and w_j. (Note that, in this subsection, we do not make any assumptions concerning the dimensionality b_3 of the vectors w_j, F and H.)

Any sequence of ISD vacua must, of course, still have a finite constant value for the quantity in Eq. (6.43), and each of the terms in that equation must thus be finite. This time however, we assume that the eigenvalues λ_j are bounded from below, and that $\tau_2 \to \infty$. The quantities ϵ_{ij} therefore must satisfy

$$\epsilon_{ij} = (v_i^F F + v_i^H H) \cdot w_j = \mathcal{O}(1/\sqrt{\mu_i}). \tag{6.48}$$

Using the fact that the w_j, as eigenvectors of a symmetric matrix, are orthonormal, it is possible to invert the above relation to yield

$$v_i^F F + v_i^H H = \mathcal{O}(1/\sqrt{\mu_i}). \tag{6.49}$$

In the decoupling limit the eigenvectors v_i and eigenvalues μ_i are given by

$$\mu_1 = \frac{\tau_2}{2} + \frac{\tau_1^2}{\tau_2} + \ldots, \qquad \mu_2 = \frac{1}{2\tau_2} - \frac{\tau_1^2}{\tau_2^3} + \ldots, \tag{6.50}$$

$$v_1 = \begin{pmatrix} -\frac{\tau_1}{\tau_2} + \frac{\tau_1(\tau_1^2-1)}{\tau_2^4} + \ldots \\ 1 - \frac{\tau_1^2}{2\tau_2^4 + \ldots} \end{pmatrix}, \qquad v_2 = \begin{pmatrix} 1 - \frac{\tau_1^2}{2\tau_2^4} + \ldots \\ \frac{\tau_1}{\tau_2} - \frac{\tau_1(\tau_1^2-1)}{\tau_2^4} + \ldots \end{pmatrix}. \tag{6.51}$$

Therefore Eq. (6.49) implies

$$\begin{aligned} v_1^F F + v_1^H H &= \mathcal{O}(1/\tau_2^2) F + \mathcal{O}(1) H = \mathcal{O}(1/\sqrt{\tau_2}), \\ v_2^F F + v_2^H H &= \mathcal{O}(1) F + \mathcal{O}(1/\tau_2^2) H = \mathcal{O}(\sqrt{\tau_2}). \end{aligned} \tag{6.52}$$

While the second equation allows for diverging F and H, the first equation implies $H = 0$ for τ_2 large enough, thus ruling out infinite sequences of vacua in this limit.

6.3.3 Series approaching a conifold point

Another commonly occurring kind of singularity in Calabi-Yau manifolds are conifold singularities. Let us address the question whether there can be infinite sequences of ISD vacua accumulating to conifold points[2]. Consider again our one-parameter models. Around the conifold point $z = 1$ the periods have expansions

$$\begin{pmatrix} \Pi_3 \\ \Pi_2 \\ \Pi_1 \\ \Pi_0 \end{pmatrix} = \begin{pmatrix} \xi \\ c_0 + c_1 \xi + \ldots \\ b_0 + b_1 \xi + \ldots \\ \frac{\xi}{2\pi i} \log(-i\xi) + a_0 + a_1 \xi + \ldots \end{pmatrix}, \tag{6.53}$$

---

[2]Close to a conifold point warping effects are large – see [155, 156] for the functional form of the corrections to the Kähler potential – and a complete analysis should take also this into account.

where $\xi \sim (z-1)$. Computing the corresponding metric $\mathcal{G}_\xi$ produces a matrix with the leading behavior

$$\mathcal{G}_\xi \sim \begin{pmatrix} -\frac{2\pi}{\ln|\xi|} + \frac{c_{11}}{\ln^2|\xi|} & \frac{b_{12}}{\ln|\xi|} + \frac{c_{12}}{\ln^2|\xi|} & \frac{b_{13}}{\ln|\xi|} + \frac{c_{13}}{\ln^2|\xi|} & \frac{b_{14}}{\ln|\xi|} + \frac{c_{14}}{\ln^2|\xi|} \\ \cdot & a_{22} + \frac{b_{22}}{\ln|\xi|} & a_{23} + \frac{b_{23}}{\ln|\xi|} & a_{24} + \frac{b_{24}}{\ln|\xi|} \\ \cdot & \cdot & a_{33} + \frac{b_{3}}{\ln|\xi|} & a_{34} + \frac{b_{34}}{\ln|\xi|} \\ \cdot & \cdot & \cdot & d_{44}\ln|\xi| + a_{44} \end{pmatrix}, \qquad (6.54)$$

where a_{ij}, b_{ij}, c_{ij} and d_{44} are constants that are determined in terms of the expansion coefficients of the periods. The eigenvectors and eigenvalues of this matrix have the following expansions

$$\begin{aligned}
\lambda_1 &= -\frac{2\pi}{\ln|\xi|} + \mathcal{O}(\ln^{-2}|\xi|), & w_1^T &= \left[1, \mathcal{O}(\ln^{-1}|\xi|), \mathcal{O}(\ln^{-1}|\xi|), \mathcal{O}(\ln^{-1}|\xi|)\right], \\
\lambda_2 &= \ell_A + \mathcal{O}(\ln^{-1}|\xi|), & w_2^T &= \left[\mathcal{O}(\ln^{-1}|\xi|), u_1^A, u_2^A, \mathcal{O}(\ln^{-1}|\xi|)\right], \\
\lambda_3 &= \ell_B + \mathcal{O}(\ln^{-1}|\xi|), & w_3^T &= \left[\mathcal{O}(\ln^{-1}|\xi|), u_1^B, u_2^B, \mathcal{O}(\ln^{-1}|\xi|)\right], \\
\lambda_4 &= d_{44}\ln|\xi| + \mathcal{O}(1), & w_4^T &= \left[\mathcal{O}(\ln^{-2}|\xi|), \mathcal{O}(\ln^{-1}|\xi|), \mathcal{O}(\ln^{-1}|\xi|), 1\right].
\end{aligned} \qquad (6.55)$$

Here $(u^{A/B})$ and $\ell_{A/B}$ are the eigenvectors and eigenvalues of the two-by-two matrix

$$\begin{pmatrix} a_{22} & a_{23} \\ a_{23} & a_{33} \end{pmatrix}, \qquad (6.56)$$

respectively. With these expansions it is straightforward to prove, in complete parallel to the LCS case, that the flux vectors $F = (F^0, F^1, F^2, F^3)$ and H must satisfy

$$F^3 = H^3 = 0, \qquad F^{1,2}, H^{1,2} = \mathcal{O}(1) \quad \text{and} \quad F^0, H^0 = \mathcal{O}(\ln^{1/2}|\xi|), \qquad (6.57)$$

as $\xi \to 0$ to be able to support an ISD vacuum. (Note that F^3 and H^3 are fluxes piercing the shrinking cycle.) At this stage, letting F^0 and H^0 go to infinity produces no contradiction. Thus, the simple argument that disproved infinite sequences in the LCS case is not sufficient for doing the same for the conifold limit. However, computing $D_\xi W$ explicitly shows that no infinite series is possible. To see this we note first that τ is given by

$$\tau = \frac{F \cdot \Pi^\dagger}{H \cdot \Pi^\dagger} = \frac{F^1 \bar{c}_0 + F^2 \bar{b}_0}{H^1 \bar{c}_0 + H^2 \bar{b}_0} + \mathcal{O}(\xi). \qquad (6.58)$$

To compute $D_\xi W$ we first record the expressions for K_ξ and $D_\xi \Pi$:

$$K_\xi = \frac{\bar{a}_0 - c_1 \bar{b}_0 + b_1 \bar{c}_0}{2i\,\mathrm{Im}\,(c_0 \bar{b}_0)} + \mathcal{O}(\xi \log \xi), \qquad D_\xi \Pi = \begin{pmatrix} 1 + \mathcal{O}(\xi) \\ A_1 + \mathcal{O}(\xi \log \xi) \\ A_2 + \mathcal{O}(\xi \log \xi) \\ \frac{1}{2\pi i}\log(-i\xi) + \mathcal{O}(1) \end{pmatrix}, \qquad (6.59)$$

with

$$A_1 = c_1 + c_0 \frac{\bar{a}_0 - c_1 \bar{b}_0 + b_1 \bar{c}_0}{2i\,\mathrm{Im}\,(c_0 \bar{b}_0)}, \qquad A_2 = b_1 + b_0 \frac{\bar{a}_0 - c_1 \bar{b}_0 + b_1 \bar{c}_0}{2i\,\mathrm{Im}\,(c_0 \bar{b}_0)}. \qquad (6.60)$$

This yields

$$\begin{aligned}
D_\xi W &= F^0 - \frac{F^1 \bar{c}_0 + F^2 \bar{b}_0}{H^1 \bar{c}_0 + H^2 \bar{b}_0} H^0 + \frac{(F^1 H^2 - F^2 H^1)\bar{a}_0}{\bar{c}_0 H^1 + \bar{b}_0 H^2} + \mathcal{O}(\xi \ln|\xi|) \\
&= F^0 - \tau H^0 + \mathcal{O}(1).
\end{aligned} \qquad (6.61)$$

We see from this expressions that in order to have $D_\xi W = 0$ as $F^0, H^0 \to \infty$, τ must approach the real ratio F^0/H^0. This means that the imaginary part of τ goes to zero, which is S-dual to a decoupling limit. Therefore, as in the LCS case, there are no infinite sequences of vacua with finite (and nonzero) string coupling.

6.3.4 The two-parameter model $\mathcal{M}_{(86,2)}$

Until now we have studied D-limits in the complex structure and axio-dilaton moduli spaces of a family of one-parameter Calabi-Yau manifolds. In this section, as a first step to a more general result, we extend the previous result to a specific two-parameter model. Again we find that there is no infinite sequence of supersymmetric vacua approaching the LCS point.

Consider the two-parameter model $\mathcal{M}_{(86,2)}$. Its periods can be expanded around the LCS point [138]:

$$\begin{pmatrix} \Pi_5 \\ \Pi_4 \\ \Pi_3 \\ \Pi_2 \\ \Pi_1 \\ \Pi_0 \end{pmatrix} \sim \begin{pmatrix} \delta_3 - \frac{25}{12} y - x - \frac{1}{6}(5y^3 + 12y^2 x) \\ -\frac{25}{12} + \frac{1}{2}y + \frac{5}{2}y^2 + 4xy \\ -1 + 2y^2 \\ y \\ x \\ 1 \end{pmatrix}. \tag{6.62}$$

Here $\delta_3 = \frac{21 i \zeta(3)}{\pi^3}$ is a constant, whereas x and y are the two complex structure moduli. Approaching the LCS point corresponds to sending $\operatorname{Im} x \to \infty$ and $\operatorname{Im} y \to \infty$, where the limits can be taken independently. The Kähler potential takes the form

$$e^{-K} = 16 \, x_2 \, y_2^2 + \frac{20}{3} y_2^3 + \frac{42 \, \zeta(3)}{\pi^3} + \ldots, \tag{6.63}$$

where $x = x_1 + i x_2$ and $y = y_1 + i y_2$ with $x_i, y_i \in \mathbb{R}$.

Consider the case in which the limits for the two variables are taken at the same time, i.e. $x_2 = y_2 = z \to \infty$. This choice significantly simplifies the computation of the metric $\mathcal{G}_z$, which results in

$$\mathcal{G}_z = \begin{pmatrix} a_{11} z^3 + \mathcal{O}(z) & a_{12} z + \mathcal{O}(\frac{1}{z}) & a_{13} z + \mathcal{O}(\frac{1}{z}) & a_{14} \frac{1}{z} + \mathcal{O}(\frac{1}{z^3}) & a_{15} \frac{1}{z} + \mathcal{O}(\frac{1}{z^2}) & a_{16} \frac{1}{z^3} + \mathcal{O}(\frac{1}{z^4}) \\ \cdot & a_{22} z + \mathcal{O}(\frac{1}{z}) & a_{23} z + \mathcal{O}(\frac{1}{z}) & a_{24} \frac{1}{z} + \mathcal{O}(\frac{1}{z^3}) & a_{25} \frac{1}{z} + \mathcal{O}(\frac{1}{z^3}) & a_{26} \frac{1}{z^3} + \mathcal{O}(\frac{1}{z^4}) \\ \cdot & \cdot & a_{33} z + \mathcal{O}(\frac{1}{z}) & a_{34} \frac{1}{z} + \mathcal{O}(\frac{1}{z^3}) & a_{35} \frac{1}{z} + \mathcal{O}(\frac{1}{z^3}) & a_{36} \frac{1}{z^3} + \mathcal{O}(\frac{1}{z^4}) \\ \cdot & \cdot & \cdot & a_{44} \frac{1}{z} + \mathcal{O}(\frac{1}{z^3}) & a_{45} \frac{1}{z} + \mathcal{O}(\frac{1}{z^3}) & a_{46} \frac{1}{z^3} + \mathcal{O}(\frac{1}{z^6}) \\ \cdot & \cdot & \cdot & \cdot & a_{55} \frac{1}{z} + \mathcal{O}(\frac{1}{z^3}) & a_{56} \frac{1}{z^3} + \mathcal{O}(\frac{1}{z^6}) \\ \cdot & \cdot & \cdot & \cdot & \cdot & a_{66} \frac{1}{z^3} + \mathcal{O}(\frac{1}{z^6}) \end{pmatrix}. \tag{6.64}$$

The constants a_{ij} are known, and we collect them in Appendix C.1.2. The eigenvectors and eigenvalues of this metric expanded in z are given by

$$\lambda_1 = a_{11}z^3 + \mathcal{O}(z), \quad w_1^T = \left[1, \mathcal{O}(z^{-2}), \mathcal{O}(z^{-2}), \mathcal{O}(z^{-4}), \mathcal{O}(z^{-4}), \mathcal{O}(z^{-6})\right],$$
$$\lambda_2 = \ell_2 z + \mathcal{O}(z^{-1}), \quad w_2^T = \left[\mathcal{O}(z^{-2}), w_2^2 + \mathcal{O}(z^{-2}), w_2^3 + \mathcal{O}(z^{-2}), \mathcal{O}(z^{-2}), \mathcal{O}(z^{-2}), \mathcal{O}(z^{-4})\right],$$
$$\lambda_3 = \ell_3 z + \mathcal{O}(z^{-1}), \quad w_3^T = \left[\mathcal{O}(z^{-2}), w_3^2 + \mathcal{O}(z^{-2}), w_3^3 + \mathcal{O}(z^{-2}), \mathcal{O}(z^{-2}), \mathcal{O}(z^{-2}), \mathcal{O}(z^{-4})\right],$$
$$\lambda_4 = \frac{\ell_4}{z} + \mathcal{O}(z^{-3}), \quad w_4^T = \left[\mathcal{O}(z^{-4}), \mathcal{O}(z^{-2}), \mathcal{O}(z^{-2}), w_4^4 + \mathcal{O}(z^{-2}), w_4^5 + \mathcal{O}(z^{-2}), \mathcal{O}(z^{-2})\right],$$
$$\lambda_5 = \frac{\ell_5}{z} + \mathcal{O}(z^{-3}), \quad w_5^T = \left[\mathcal{O}(z^{-4}), \mathcal{O}(z^{-2}), \mathcal{O}(z^{-2}), w_5^4 + \mathcal{O}(z^{-2}), w_5^5 + \mathcal{O}(z^{-2}), \mathcal{O}(z^{-2})\right],$$
$$\lambda_6 = \frac{a_{66}}{z^3} + \mathcal{O}(z^{-6}), \quad w_6^T = \left[\mathcal{O}(z^{-6}), \mathcal{O}(z^{-4}), \mathcal{O}(z^{-4}), \mathcal{O}(z^{-2}), \mathcal{O}(z^{-2}), 1\right]. \quad (6.65)$$

The ℓ_i and w_i^j are the eigenvalues and eigenvectors of appropriate two-by-two matrices. Consider the flux-vector $N_n = (N_n^0, \ldots, N_n^5)$ and the following limit:

$$\lim_{n \to \infty} z_n = \infty, \quad \lim_{n \to \infty} N_n \cdot \mathcal{G}_{z_n} \cdot N_n^T \equiv \lim_{z \to \infty} \sum_{i=1}^{6} |N \cdot w_i(z)|^2 \lambda_i(z) \neq \infty. \quad (6.66)$$

In order not to clutter notation, we will suppress the index n in the following.

Without loss of generality assume $N^0 = 1$. The first eigenvalue grows as $\lambda_1 \sim z^3$. Therefore we must have $|N \cdot w_1|^2 \sim \mathcal{O}(z^{-3})$. Hence

$$N \cdot w_1 \sim \mathcal{O}(z^{-3/2}) = 1 + N^1 \mathcal{O}(z^{-2}) + N^2 \mathcal{O}(z^{-2}) + N^3 \mathcal{O}(z^{-4}) + N^4 \mathcal{O}(z^{-4}) + N^5 \mathcal{O}(z^{-6}). \quad (6.67)$$

This can happen only if at least one of the N^i diverges. It is also clear that this must happen in order for N to approach one of the zero eigenvectors of $\mathcal{G}_z$. What is needed is

$$N^1 = Pz^2 + o(z^2), \quad N^2 = Qz^2 + o(z^2),$$
$$N^3 = Rz^4 + o(z^4), \quad N^4 = Sz^4 + o(z^4),$$
$$N^5 = Tz^6 + o(z^6). \quad (6.68)$$

Recall that $o(z^2)$ stands for terms that grow slower than z^2 and P, Q, R, S, T are appropriate constants.

Consider

$$\mathcal{O}(z^{3/2}) = N \cdot w_6 = Tz^6 + S\mathcal{O}(z^2) + R\mathcal{O}(z^2) + \ldots \quad (6.69)$$

This immediately proves that T must be zero. We set it to zero in the following. Furthermore, consider

$$\mathcal{O}(z^{1/2}) = N \cdot w_4 = (Rw_4^4 + Sw_4^5)z^4 + \ldots, \quad (6.70)$$

and

$$\mathcal{O}(z^{1/2}) = N \cdot w_5 = (Rw_5^4 + Sw_5^5)z^4 + \ldots \quad (6.71)$$

This proves that $(Rw_4^4 + Sw_4^5) = (Rw_5^4 + Sw_5^5) = 0$, i.e.

$$\begin{pmatrix} w_4^4 & w_4^5 \\ w_5^4 & w_5^5 \end{pmatrix} \begin{pmatrix} R \\ S \end{pmatrix} = 0. \quad (6.72)$$

Since the matrix is orthogonal it follows that $R = S = 0$.

We continue the analysis along the line of the LCS case of the one-parameter models. In the end we obtain following conditions on the flux-vectors $F = (F^0, \ldots, F^5)$ and $H = (H^0, \ldots, H^5)$:

$$F^0 = F^1 = F^2 = 0 \quad \text{and} \quad H^0 = H^1 = H^2 = 0 \,. \tag{6.73}$$

This result means that there is no ISD vacuum approaching the LCS. The only exception is again the singular vacuum located exactly at the LCS point.

6.4 D-limits and infinite flux series for F-theory on $K3 \times K3$

The simplest non-trivial flux compactifications apart from toroidal orbifolds are compactifications of type IIB string theory on the orientifold $K3 \times T^2/\mathbb{Z}_2$. These models contain four orientifold planes and 16 D7-branes which are points in $T^2/\mathbb{Z}_2$ and fill out the entire $K3$ as well as the four non-compact directions.

Alternatively, these compactifications can be described as F-theory on $K3 \times K3$. This description not only allows for an elegant treatment of IIB flux compactifications on the orientifold $K3 \times T^2/\mathbb{Z}_2$, but also naturally includes two-form fluxes on the D7-branes. As shown by Aspinwall and Kalloch [151], the number of supersymmetric vacua of such compactifications is finite, i.e. there can be no infinite flux series in these models.

In this section, we discuss this result from the perspective of D-limits. We are able to find D-limits as well as associate infinite flux sequences on $K3$, so that the result of [151] implies that all but finitely many of the corresponding solutions are actually equivalent by automorphism of the lattice $H^2(K3, \mathbb{Z})$. We demonstrate this in a simple example.

Compactifications of type IIB string theory on $K3 \times T^2/\mathbb{Z}_2$ with G_3 flux have also been considered in [157]. They show how to find an infinite sequence of fluxes which solves all of the supersymmetry conditions except for primitivity. In general, imaginary self-duality (ISD) does not imply supersymmetry. In the present case, however, one can show that the complex structure of $K3$ may always be chosen such that (for fixed metric) the supersymmetry constraints are satisfied for any ISD solution. Hence their sequence also breaks imaginary self-duality. Therefore, we can not treat the flux series of [157] as a D-limit in the sense introduced.

6.4.1 F-theory with $G_{(4)}$ flux on $K3 \times K3$

In compactifications of F-theory on the fourfold $K3 \times K3$ [3], one can switch on four-form fluxes $G_{(4)}$ which are integrally quantized. They can be written as

$$G_{(4)} = G^{\mu\nu} \eta_\mu \wedge \tilde{\eta}_\nu \,. \tag{6.74}$$

---

[3]One of the $K3$s has to be elliptically fibered for F-theory to make sense. We assign tildes to quantities associated with the elliptic $K3$.

Here η_μ and $\tilde\eta_\nu$ are integral two-forms on the two $K3$s. We will think of the matrix $G^{\mu\nu}$ as the components of a vector in $H^2(K3,\mathbb{Z}) \otimes H^2(K3,\mathbb{Z})$ and simply write G in the following.

The scalar potential induced by the fluxes can stabilize both complex structure as well as Kähler moduli of $K3 \times K3$ (except for the volumes of the two $K3$s). The vacua of these models were analyzes in [151, 158]. See also [157, 159] for an analysis from the type IIB perspective.

It can be shown that the scalar potential is positive definite and can be written as [158]

$$V = \frac{1}{2} \int_{K3\times K3} G_{(4)} \wedge (*G_{(4)} - G_{(4)}). \tag{6.75}$$

As $G_{(4)}$ is forced to be self-dual by the equations of motion, their solutions correspond to Minkowski minima of the effective potential.

For $K3 \times K3$, the tadpole condition (6.26) reads

$$\frac{1}{2} \int_{Y_4} G_{(4)} \wedge G_{(4)} + N_{D3} = 24, \tag{6.76}$$

which can be rewritten as

$$24 = N_{D3} + \frac{1}{2}\int_{K3\times K3} G_{(4)} \wedge G_{(4)} = N_{D3} + \frac{1}{2}\int_{K3\times K3} G_{(4)} \wedge *G_{(4)}, \tag{6.77}$$

which is manifestly positive. In order to discuss D-limits, we consider the metric $\mathcal{G}$, which is defined by

$$G \cdot \mathcal{G} \cdot G^T \equiv \frac{1}{2}\int_{K3\times K3} G_{(4)} \wedge *G_{(4)}. \tag{6.78}$$

As we consider a fourfold which is a product of two spaces, we can decompose

$$\mathcal{G} = \frac{1}{2}\mathcal{G}_\Sigma \otimes \mathcal{G}_{\tilde\Sigma}. \tag{6.79}$$

Given an integral two-form $G_{(2)} = \sum_\mu g_\mu \eta^\mu$, $\mathcal{G}_\Sigma$ is defined by

$$\int_{K3} G_{(2)} \wedge *G_{(2)} = g \cdot \mathcal{G}_\Sigma \cdot g^T. \tag{6.80}$$

As before, a D-limit is defined to be a limit in moduli space in which $\mathcal{G}$, i.e. $\mathcal{G}_\Sigma$ or $\mathcal{G}_{\tilde\Sigma}$, degenerates.

6.4.2 The $K3$ surface

In order to discuss the properties of $\mathcal{G}_\Sigma$ and find which D-limits we can have for F-theory compactification on $K3 \times K3$ we collect a few crucial properties about $K3$ surfaces in this section. For a more thorough treatment, see e.g. [160, 161].

In two complex dimensions there is just one non-trivial compact Calabi-Yau manifold: $K3$. The metric deriving from the natural inner product on the 22-dimensional space $H^2(K3)$:

$$M_{\mu\nu} = \int_{K3} \eta_\mu \wedge \eta_\nu, \tag{6.81}$$

has signature (3, 19). The vector space $H^2(K3)$ contains the lattice $H^2(K3, \mathbb{Z})$, the elements of which are Poincaré dual to curves in $K3$. This lattice can be written as

$$H^2(K3, \mathbb{Z}) = -E_8^{\oplus 2} \oplus U^{\oplus 3}, \qquad (6.82)$$

where E_8 denotes the root lattice of E_8 and U is the hyperbolic lattice. Embedded in a vector space with orthonormal basis E_I, the root lattice of E_8 is given by vectors

$$\sum_I q_I E_I, \qquad (6.83)$$

where the q_I have to be *all* integer or *all* half-integer and fulfill the relations $\sum_{I=1,\ldots,8} q_I \in 2\mathbb{Z}$. The lattice $U^{\oplus 3}$ is spanned by integral multiples of e_i, e^j which have the intersections

$$e_i \cdot e^j = \delta_i^j, \qquad e_i \cdot e_j = 0, \qquad e^i \cdot e^j = 0. \qquad (6.84)$$

The inner product between integral two-forms has a geometric interpretation as the intersection of the dual curves. As the $K3$ surface has a trivial canonical bundle, the self-intersection number of a curve, i.e. the intersection between two homologous curves, translates to its genus by using the adjunction formula. Denoting the curve dual to the integral two-form η_C by C one obtains

$$\int_{K3} \eta_C \wedge \eta_C = C \cap C = -\chi(C) = -2 + 2g(C). \qquad (6.85)$$

The geometric moduli space of $K3$ is the set of all oriented positive-norm three-planes Σ in $H^2(K3)$ modulo automorphisms of the lattice $H^2(K3, \mathbb{Z})$ in $O^+(3, 19)$ [160, 162–164]. The group $O^+(3, 19)$ is the component of the orthogonal group which leaves the orientation of Σ invariant.

We span Σ using three orthonormal vectors ω_i:

$$\omega_i \cdot \omega_j = \delta_{ij}. \qquad (6.86)$$

Note that this description leaves an $SO(3)$ symmetry, rotating the ω_i into one another. We can construct the Kähler form J and the holomorphic two-form Ω of $K3$ using the vectors ω_i:

$$\Omega = \omega_1 + i\omega_2, \qquad J = \sqrt{V}\omega_3, \qquad (6.87)$$

where we have denoted the volume of $K3$ by V. It is important to note that a choice of Σ determines the metric of $K3$ (up to the overall volume), but does not completely determine the complex structure. We still may rotate the ω_i inside Σ or equivalently change the definition in (6.87). For a fixed complex structure, the lattice of integral cycles of $K3$ which are orthogonal to Ω is the Picard lattice.

Any two-form $H \in H^2(K3)$ can be decomposed into a piece parallel and a piece perpendicular to Σ:

$$H = H_\parallel + H_\perp. \qquad (6.88)$$

The action of the Hodge-$*$ operation on $K3$ then takes the simple form

$$*H = *H_\parallel + *H_\perp = H_\parallel - H_\perp. \qquad (6.89)$$

The $K3$ moduli space naturally includes loci over which the $K3$ surface develops ADE singularities. Whenever there are elements $\gamma_i \in H^2(K3, \mathbb{Z})$ with $\gamma_i \cdot \gamma_i = -2$ that are orthogonal to Σ, the dual spheres collapse to produce an ADE singularity. Loci where this occurs are at a finite distance in moduli space from any generic smooth $K3$.

6.4.3 D-limits and $\mathcal{G}_\Sigma$

Let us now see if $\mathcal{G}_\Sigma$ can degenerate so that we find a D-limit. As before, we denote the vector of coefficients that is obtained when an integral two-form $G_{(2)}$ is expanded in some basis η_μ of $H^2(K3)$ by g. In order to facilitate the discussion of flux quantization we choose this basis to be integral, i.e. the vectors η_μ are elements of the lattice $H^2(K3, \mathbb{Z})$.

Using this basis,
$$\omega_i = \omega_i^\mu \eta_\mu, \qquad G_{(2)} = g^\mu \eta_\mu. \tag{6.90}$$

We can decompose
$$G_{(2)} = G_{(2)\|} + G_{(2)\perp} = \sum_i \int_{K3} (\omega_i \wedge G_{(2)})\, \omega_i + G_{(2)} - \sum_i \int_{K3} (\omega_i \wedge G_{(2)}) \omega_i. \tag{6.91}$$

Hence
$$*G_{(2)} = G_{(2)\|} - G_{(2)\perp} = \left(2 M_{\mu\nu} \sum_i \omega_i^\rho \omega_i^\mu - \delta_\nu^\rho \right) g^\nu \eta_\rho. \tag{6.92}$$

Defining the projector
$$\Gamma_\nu^\rho \equiv \sum_i \omega_i^\rho \omega_i^\mu M_{\mu\nu}, \qquad \Gamma^2 = \Gamma, \tag{6.93}$$
which projects any form onto its components parallel to Σ, we obtain
$$\mathcal{G}_\Sigma = M\,(2\Gamma - 1). \tag{6.94}$$

As the inner product (6.80) is positive definite for any smooth $K3$ surface, it follows that the metric in (6.94) has the same property. It can only degenerate in a limit in moduli space in which the $K3$ surface becomes singular. Let us first consider the aforementioned ADE singularities. They occur when we rotate the three-plane Σ such that it becomes orthogonal to specific lattice vectors of $H^2(K3, \mathbb{Z})$ with respect to the metric (6.81). The expression we have derived for $\mathcal{G}_\Sigma$, however, does not at all depend on the location of Σ relative to the lattice $H^2(K3, \mathbb{Z})$. Hence the metric $\mathcal{G}_\Sigma$ can not degenerate when we approach a locus in moduli space for which the $K3$ surface has an ADE singularity. Note also that these singularities occur at finite distance in moduli space, i.e. the naturally lie *inside* the moduli space. Another kind of singularity occurs when we rotate Σ towards a light-like direction in $H^2(K3)$. In the following, we shall investigate such a limit and show that it indeed gives rise to a degeneration of $\mathcal{G}_\Sigma$.

A well-known example of such a limit is the F-theory limit of a compactification of M-theory on an elliptically fibered $K3$. In this limit, the volume of the T^2 fiber is taken to zero, which corresponds to rotating the Kähler form towards the light cone in $H^2(K3)$ [165]. Just as in the case of the large complex structure limit, this limit is dual to a decompactification limit which takes place on the type IIB/F-theory side. Furthermore, it can be shown that this limit is at infinite distance in moduli space [158].

An example

To show that rotating Σ towards the light cone constitutes a D-limit, we consider a simple example. For ease of exposition, we keep the three-plane Σ in a four-dimensional subspace spanned by

$$d^1 = e_1 + e^1, \qquad d^2 = e_2 + e^2, \qquad d^3 = e_3 + e^3, \qquad d^4 = e_1 - e^1. \qquad (6.95)$$

Note that the intersection form is diagonal in terms of the d^i: $M = \text{diag}(2, 2, 2, -2)$. In this basis, we choose Σ to be spanned by the orthonormal vectors

$$\omega_1(n) = \frac{1}{\sqrt{2}}(1, 0, n, n), \qquad \omega_2(n) = \frac{1}{\sqrt{2}}(0, 1, 0, 0), \qquad \omega_3(n) = \frac{1}{\sqrt{2 + 2n^2}}(n, 0, -1, 0). \qquad (6.96)$$

The matrix $\mathcal{G}_\Sigma = M(2\Gamma - 1)$ is

$$\mathcal{G}_\Sigma = 2 \begin{pmatrix} \frac{1+3n^2}{1+n^2} & 0 & \frac{2n^2}{1+n^2} & -2n \\ 0 & 1 & 0 & 0 \\ \frac{2n^2}{1+n^2} & 0 & -1 + 2(n^2 + \frac{1}{1+n^2}) & -2n^2 \\ -2n & 0 & -2n^2 & 1 + 2n^2 \end{pmatrix}. \qquad (6.97)$$

Its eigenvalues are given by

$$\lambda_1 = \lambda_2 = 2, \qquad \lambda_3 = 2 + 4n^2 - 4\sqrt{n^2 + n^4}, \qquad \lambda_4 = 2 + 4n^2 + 4\sqrt{n^2 + n^4}. \qquad (6.98)$$

In the limit $n \to \infty$ we may approximate

$$\sqrt{n^2 + n^4} = n^2 + \frac{1}{2} - \frac{1}{8n^2} + \mathcal{O}(n^{-4}). \qquad (6.99)$$

In this limit we hence find that λ_3 goes to zero and λ_4 goes to infinity:

$$\lambda_3 \sim \frac{1}{2n^2}, \qquad \lambda_4 \sim 8n^2. \qquad (6.100)$$

The eigenvector v_3 associated with λ_3 is

$$v_3 = \begin{pmatrix} \frac{-n^3 + n(-1+\sqrt{n^2+n^4})}{(1+n^2)(n^2-\sqrt{n^2+n^4})} \\ 0 \\ -\frac{n^2(1+n^2-\sqrt{n^2+n^4})}{(1+n^2)(n^2-\sqrt{n^2+n^4})} \\ 1 \end{pmatrix} \xrightarrow{n \to \infty} \begin{pmatrix} n^{-1} \\ 0 \\ 1 \\ 1 \end{pmatrix}. \qquad (6.101)$$

Hence we find a D-limit in which the metric degenerates in the direction of ω_1. Note that this is precisely the direction of Σ which we are rotating towards the light-cone.

Let us now use this example to construct a flux series. The flux vector

$$g = (1, 0, n, n, 0, 0, ...) \sim nv_3 \,, \tag{6.102}$$

is properly quantized for any integer n. For large values of n we have that

$$g \cdot \mathcal{G}_\Sigma \cdot g^T \sim n^2 \frac{1}{n^2} = \text{const} \,. \tag{6.103}$$

Hence the eigenvalue of v_3 goes to zero fast enough to allow for an infinite sequence of integral flux vectors for which $g \cdot \mathcal{G}_\Sigma \cdot g^T$ approaches a constant in the D-limit.

6.4.4 Infinite series and automorphisms of $H^2(K3, \mathbb{Z})$

To put the example of the last section to work we set

$$\Omega(n) = \omega_1(n) + i\omega_2 \,, \tag{6.104}$$

$$\tilde{\Omega} = \tilde{\omega}_1 + i\tilde{\omega}_2 \,, \tag{6.105}$$

with $\omega_i(n)$ given by (6.96). A properly quantized flux series that obeys the supersymmetry conditions (and equations of motion) is given by [151]

$$G_{(4)}(n) = \sqrt{2} \left(\omega_1(n) \wedge \tilde{\omega}_1 + \omega_2 \wedge \tilde{\omega}_2 \right) . \tag{6.106}$$

The flux-induced D3 tadpole is

$$\frac{1}{2} \int_{K3 \times K3} G_{(4)}(n) \wedge G_{(4)}(n) = 2 \,, \tag{6.107}$$

for any n. For the Kähler form J we can choose any positive norm two-form in $H^2(K3)$ which is orthogonal to Ω. Setting $J = \omega_3(n)$ demonstrates that such a J can always be found.

In [151], it was shown that there can only be a finite number of supersymmetric flux vacua in compactifications on $K3 \times K3$. In order to make contact with our results, we review their main results. As supersymmetry demands that the flux is of type $(2, 2)$ and primitive, one can write

$$G = \text{Re}\left(c\Omega \wedge \bar{\tilde{\Omega}}\right) + \sum_\alpha \psi_\alpha \wedge \tilde{\psi}_\alpha \,, \tag{6.108}$$

where c is a parameter that has to be chosen appropriately for flux quantization and $\psi_\alpha, \tilde{\psi}_\alpha$ are integral primitive $(1, 1)$ forms on the respective $K3$ surfaces. They show that if only the first term is present, as is the case for our example, the complex structure moduli of the two $K3$ surfaces, i.e. Ω and $\tilde{\Omega}$, are completely fixed. Furthermore, they are fixed such that the Picard lattice of the corresponding $K3$ surfaces is of maximal rank, i.e. Ω sits inside a two-dimensional lattice $\Upsilon \subset H^2(K3, \mathbb{Z})$. Such $K3$ surfaces, which have been dubbed[4] 'attractive', can be classified through the lattice Υ. It turns out that only a finite number of attractive

---
[4]They have also been referred to as 'singular' $K3$ surfaces, even though they can be perfectly smooth manifolds. Hence we follow [151] in calling them 'attractive'.

$K3$ surfaces can satisfy the tadpole condition (6.76). When the second term in (6.108) is also present, the $K3$ ceases to be attractive. Its contribution to the tadpole is, however, always positive definite. Hence there can be only a finite number of flux choices that admit supersymmetric flux vacua and satisfy the tadpole condition for F-theory on $K3 \times K3$.

Supersymmetry only forces $G_{(4)}$ to be primitive, but does not fix the Kähler moduli. Non-perturbative effects, however, give rise to an effective potential that can fix all Kähler moduli. As the effective potential is determined once fluxes (and hence the complex structure) are given, it follows that there is only a finite number of supersymmetric stable flux vacua for F-theory on $K3 \times K3$. In case both terms in (6.108) are non-zero, some of the instantons that stabilize the Kähler moduli can be obstructed, so that not all moduli are stabilized.

The results of [151] indicate that all but a finite number of the vacua of the series we have constructed before must actually be equivalent. Note that for our series, only the first term in (6.108) is present. Once we specify G in terms of Ω, the Kähler form is therefore determined completely. Hence we have to show that there is an automorphism of $H(K3, \mathbb{Z})$ which identifies all but a finite number of the sub-lattices spanned by the $\Omega(n)$. To find this automorphism, we write $\omega_1(n)$, ω_2 in terms of a basis for the lattice $U^{\oplus 3}$:

$$\omega_1(n) = e_1(1+n) + ne_3 + e^1(1-n) + ne^3, \qquad (6.109)$$

$$\omega_2 = e_2 + e^2. \qquad (6.110)$$

Indeed, there is an automorphism of $H^2(K3, \mathbb{Z})$ which identifies *all* of the solutions in our series. It is given by

$$\begin{aligned} e_1 &\mapsto \hat{e}_1 = (1+n)e_1 + ne_3, & e^1 &\mapsto \hat{e}^1 = (1-n)e^1 + ne^3, \\ e_2 &\mapsto \hat{e}_2 = -e_2, & e^2 &\mapsto \hat{e}^2 = -e^2, \\ e_3 &\mapsto \hat{e}_3 = ne_1 + (n-1)e_3, & e^3 &\mapsto \hat{e}^3 = ne^1 - (1+n)e^3, \end{aligned} \qquad (6.111)$$

with all other elements unchanged. It maps

$$\omega_1(0) = e_1 + e^1 \mapsto \hat{e}_1 + \hat{e}^1 = \omega_1(n), \qquad (6.112)$$

$$\omega_2 \mapsto -\omega_2. \qquad (6.113)$$

Hence this automorphism identifies the holomorphic two-forms and consequently also the fluxes of our series of $K3$ surfaces. Furthermore, it gives rise to an orientation preserving[5] map of Σ to itself. Thus it is induced from a diffeomorphism of $K3$ [162–164], so that all of the solutions in our series should be considered equivalent.

Our example is, of course, very simple in that it only rotates Ω towards the light cone in the lattice $U^{\oplus 3}$. Even though examples of D-limits and infinite flux series employing the E_8 lattices can be constructed in a straightforward fashion, the corresponding automorphisms are harder to find. Showing that such automorphisms exist for any D-limit would hence constitute an alternative proof of the finiteness of the number of supersymmetric flux vacua on $K3 \times K3$.

---
[5]Note that this is not the case if we leave e_2 and e^2 invariant.

As the self-duality condition on G_4 follows from the equations of motion but does not require supersymmetry, one could then try to prove a similar theorem also for non-supersymmetric vacua.

One can turn this logic around and construct automorphism of $K3$ by studying D-limits. By the result of [151], only a finite number of solutions in any infinite sequence of supersymmetric vacua can be different. Hence there must be corresponding automorphisms in $O^+(3,19)$ which identify all but a finite number of the solutions. It would be interesting to use this approach to study the diffeomorphism group of $K3$ surfaces.

As the self-duality condition also holds without supersymmetry, infinite sequences can also only occur in D-limits in this case. With a sufficient understanding of the automorphisms of $K3$ it hence seems possible to use the D-limit approach to study the existence of infinite sequences of non-supersymmetric solutions.

6.5 The models of Ahlqvist et al.

In this section we take a closer look at a few examples of sequences of minima that converge to the LCS point, and that were first reported on in [141]. These minima have vanishing scalar potential and hence fulfill the ISD condition. A question left open in this reference was whether these series are infinite. Here we use the LCS expansions of the periods to show that there are more minima in the series than those reported in [141], but that the minima eventually break the ISD condition and the series terminate in agreement with the discussion in section 6.3.1. After a brief description of the method we used to find the minima, we present two examples of sequences of minima.

To speed up the numerical calculation of the potential, we proceed as follows. We first compute the periods and their derivatives on a grid in the complex structure modulus plane. This computation is performed using the built-in Meijer functions of Maple for the full periods, and using Matlab for the LCS expansions of the periods. We then feed these periods into Matlab where the superpotential, Kähler and scalar potentials are computed. We also use Matlab to find the minima of the potential, and determine their position and minimum value of the potential.

Since the minima in the series approach the LCS point, the LCS expansion of the periods provides a good and computationally cheap approximation of the features of the minima closest to this point. An illustration of this is shown in the figures 6.1 and 6.2, where the Mirror Quintic potential for the flux configuration $H = (-2, -4, -33, 0)$ and $F = (3, -18, 9, -1)$ is plotted using both the full Meijer functions and the LCS expansions. As can be seen from the figures, the two potentials are very similar; in particular the location and value of the potential in the minimum agree to a good degree. Consequently, the LCS expansions determine the features of minima to a good approximation at least for $|z| < 0.2$.

Given that the periods are computed on a grid, the position of a minimum of the potential can never be determined to a better accuracy than the grid spacing. Thus minima that lie

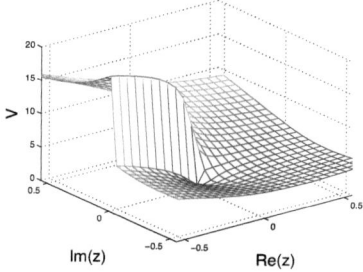 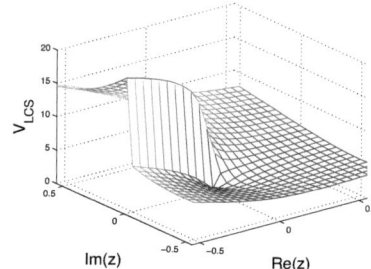

Figure 6.1: The scalar potential for the complex structure modulus z of the Mirror Quintic with NS-NS flux $H = (-2, -4, -33, 0)$ and R-R flux $F = (3, -18, 9, -1)$. The potential has already been minimized with respect to the axio-dilaton τ, so the minimum shown is a minimum for both z and τ. The first panel shows the potential calculated with periods calculated with the full Meijer functions, whereas the second potential is calculated with the LCS expansions of the periods. As can be seen from the figures, the LCS expansions are enough to reproduce the features of this minimum.

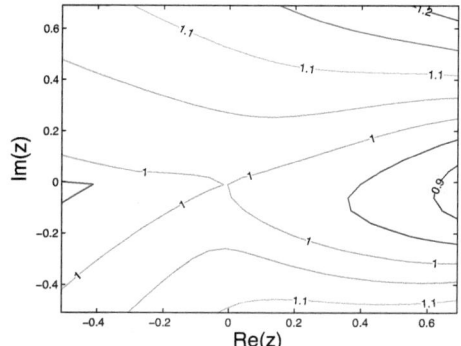

Figure 6.2: The ratio of the scalar potential computed with the Meijer functions and the potential computed using the LCS expansions, for the same fluxes as the previous figure. The closer to the LCS point, the better the match between the two potentials.

closer to the LCS point remain undetected until the grid spacing is refined. For computationally expensive functions such as the Meijer functions, this provides a significant obstacle, in that refining the grid soon becomes practically impossible. On the other hand, the LCS expansions are simple functions that can easily be computed on more and more refined grids. In figure 6.3 we show a more detailed picture of the Mirror Quintic minimum that was obtained using the LCS expansions of the periods.

Thus, in order to investigate whether the series of minima reported on in [141] continue

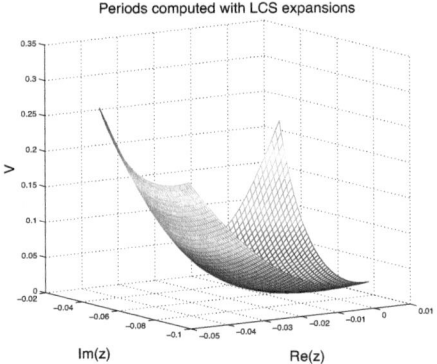

Figure 6.3: Using the LCS expansions of the periods allow us to zoom in on the supersymmetric Mirror Quintic minimum of figure 6.1.

indefinitely, we use the LCS expansion of the periods. We first compute the potential for a flux configuration on a sparse grid, identifying the region in the z-plane where the minimum is located. At this stage, we also note if we need to move the branch cut that emerges from the LCS point in order to trace the minimum to another level in the potential.[6] We then zoom in on the region that should contain a minimum and recompute the potential on a narrow grid around this point. This allows us to compute the location and potential value of the minimum to a higher accuracy. We then act on the flux vectors with the conifold monodromy matrices, and repeat the calculations for the next minimum in the series.

A series of minima on the Mirror Quintic

Using the outlined procedure, we reproduce the minima with $F^0 = 3...9$ in the Mirror Quintic series reported on in table 3 and figure 5 of [141]. In addition we find new minima with $F^0 = -17... - 6$ and $-3...2$. We found no minima for the two values $F^0 = -5, -4$, despite having studied the downward spiral of the scalar potential until it reaches its lowest level and turns back up. The z-distribution of the minima in the series is shown in figure 6.4. As can be seen, starting from $F^0 = 9$ the series of minima approaches the LCS point for decreasing values of F^0. However, as the by now negative F^0 increases in magnitude, the minima again recede from the LCS point, until they leave the region where the LCS expansion can be trusted. Thus, this series is not infinite.

As shown in figure 6.5, all minima with positive F^0 have vanishing potential in the minimum, and fulfill the ISD condition. Conversely, the minima with negative F^0 have a non-zero potential value. Thereby, this example confirms our general result that the series of minima that converge to the LCS point eventually break the ISD condition, thus inducing non-zero F-terms also in

---

[6]In some cases, it is necessary to move several steps down in the potential spiral to find the minimum, and for some flux values no minimum is found, even at the lowest level of the potential.

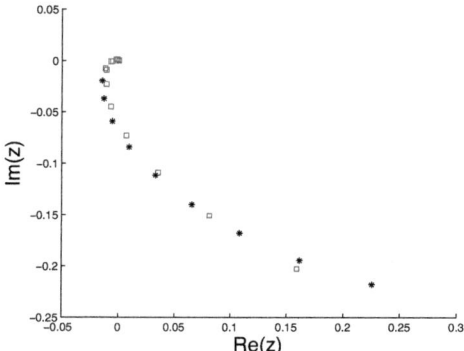

Figure 6.4: The distribution in the z plane of a series of minima that approach the large complex structure point in the Mirror Quintic moduli space. The minima have NS-NS flux $H = (-2, -4, -33, 0)$ and R-R flux $F = (F^0, -18, 9, -1)$, where F^0 ranges from -17 to 9. The red squares indicate minima with negative F^0, whereas black stars are used for minima with positive F^0.

the complex structure and axio-dilaton directions.

Figure 6.5 also shows the vacuum expectation value for the superpotential for the series of minima. Since this is large for all minima, supersymmetry is broken by the Kähler moduli, which have non-zero F-terms. We note that the tadpole for this series of minima is high, so the phenomenological interest of these minima is fairly limited.

From figure 6.5 we also see that $\text{Im}\tau$ does not run away, but stays in the range $4 - 8$. Consequently, $\mathcal{G}_\tau$ does not degenerate, and therefore this series of minima does not lie in a decompactification limit of the axio-dilaton part of moduli space.

A series of minima on Model 12

The longest series of minima that was reported on in table 3 and figure 5 of [141] was found on the one-parameter Calabi-Yau known as Model 12. This series consists of twenty-nine minima, with NS-NS flux $H = (-2, -4, -33, 0)$ and R-R flux $F = (F^0, -18, 9, -1)$, $F^0 = 7, ..., 36$. Using the LCS expansions of the periods, we reproduce some minima of this series and extend it to smaller values of F^0, as shown in figure 6.6. Just as for the Mirror Quintic example, we find that more minima exist in the vicinity of the LCS point, but the minima bounce out from the LCS point again as F^0 becomes large and negative. Thus, this series of minima does not continue indefinitely.

The value of the potential, superpotential and $\text{Im}\tau$ for Model 12 are presented in figure 6.7. As can be seen, the features are similar to the Mirror Quintic series. As expected, the ISD condition is eventually broken for negative values of F^0, and $\text{Im}\tau$ stays finite for the whole series. The superpotential is large and negative also for this series, and the tadpole is the same

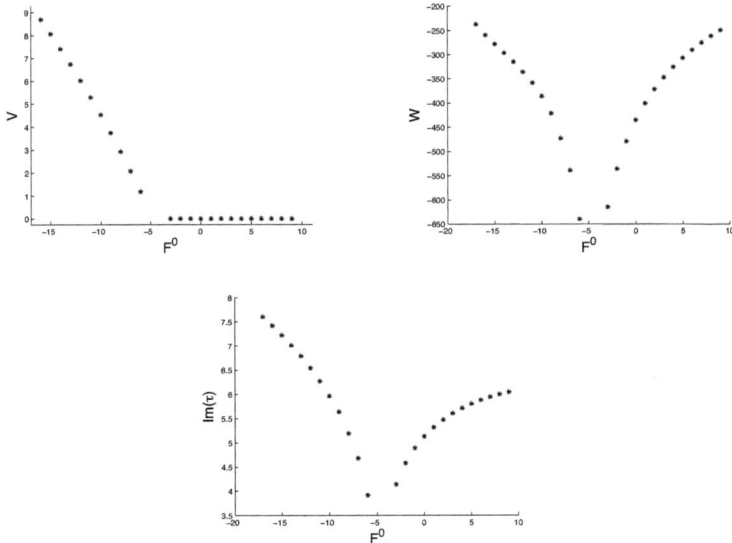

Figure 6.5: Here we show the value of the potential, the superpotential and the imaginary part of the axio-dilaton τ for the series of Mirror Quintic minima with NS-NS flux $H = (-2, -4, -33, 0)$ and R-R flux $F = (F^0, -18, 9, -1)$. As can be seen from the first panel, all minima with positive F^0 are ISD and have vanishing scalar potential. The value of the superpotential is large and negative for all minima in the series, and the dilaton does not run away to zero or infinity.

as for the Mirror Quintic series.

6.6 Summary and outlook

In this chapter we have extended the no-go result of Ashok and Douglas to include also regions around certain D-limits. For a class of one-parameter models we studied the large complex structure limit, the conifold point and the decoupling limit, and found that none of these can support infinite sequences of ISD vacua. This analysis was performed by explicitly computing a certain positive definite quadratic form defined on the space of flux quanta. This form gives the total D3-brane charge originating from three-form flux in the case of an ISD vacuum. By analysing the precise form of the eigenvalues and eigenvectors as the various D-limits are approached we demonstrated that no infinite sequences are possible. We also extended this analysis to the LCS limit of a two-parameter model, again finding that no infinite sequences exist. Furthermore, we explained how infinite sequences accumulating to D-limits in $K3 \times K3$ compactifications really correspond to finitely many vacua after the automorphism group is taken into account.

To complement the analytical results, we studied two of the sequences found by Ahlqvist et

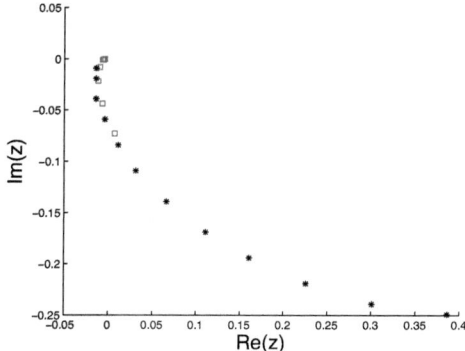

Figure 6.6: The distribution in the z plane of a series of minima that approach the large complex structure point in the moduli space of Model 12. The vacua have NS-NS flux $H = (-2, -4, -33, 0)$ and R-R flux $F = (F^0, -18, 9, -1)$, where $F^0 = -12...11$. The red squares indicate minima with negative F^0, whereas black stars are used for minima with positive F^0.

al. [141] numerically. We used expansions around the LCS point to facilitate the computations of the periods, thus making a fine grid possible. The sequences were found to turn close to the LCS point and then be repelled from it, eventually violating the ISD condition, perfectly in line with the analytical results.

In the present work we used fairly pedestrian methods to analyze the structure of the quadratic form around various singularities. For this, we needed expansions of the periods in the D-limit under consideration. This is in contrast to the statistical analysis, where the number of vacua is estimated without such detailed understanding of the Calabi-Yau. Although our method requires more information, it allows us to refine the results of the statistical analysis in the models we consider. It would of course be very interesting to formulate more general and transparent conditions on the singularity required for infinite sequences. Such a result would be a step towards a more general finiteness theorem.

Additionally, two interesting directions of future research would be to investigate whether similar techniques can be applied also in the case of generalized Calabi-Yau manifolds and to analyze how warping corrections affect the results for sequences accumulating to a conifold point.

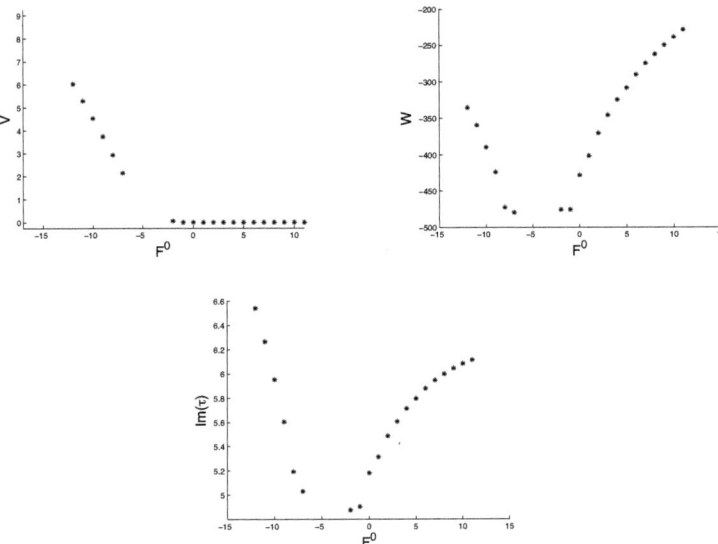

Figure 6.7: The value of the potential, the superpotential and the imaginary part of the axio-dilaton τ for the series of Model 12 minima with NS-NS flux $H = (-2, -4, -33, 0)$ and R-R flux $F = (F^0, -18, 9, -1)$ of Model 12. The features of this series of minima closely parallels those of the Mirror Quintic series.

Acknowledgements

I dedicate this work to my doctoral advisor Maximilian Kreuzer, who passed away before it was completed. Having been his student for more than three years, I have greatly profited from his vast knowledge and dedicated support.

I am thankful to Anton Rebhan for accepting to be referee of my doctoral thesis. I would like to thank Radoslav Rashkov for his support and his constant encouragement. I cordially wish to thank Harald Skarke in particular for his help in the completion of version 2.1 of PALP. It is a pleasure to thank Ralph Blumenhagen for accepting to be co-referee of my thesis, for advices and for his invitation to present some of my work's results at the MPI for Physics in Munich.

I am very thankful to all the members and friends I have met at the institute over the past four years. Especially, I would like to mention the following people: Andreas Braun for help with the manuscript, for the good music and for providing delicious food. Andrés Collinucci for collaboration and for all the South Park episodes. Niklas Johansson for help with the manuscript and for his many 'words of wisdom' always pervaded with desecrating sense of humor. Gandalf Lechner for help with the manuscript. Christoph Mayrhofer for distracting me from work when it was really needed. Michal Michalčik for providing delicious coffee and for his precious help in troubleshooting my Linux desktop.

I am in great debt with those dear friends who always have been there when I needed them the most. Grazie a te Arianna per tutti i momenti che abbiamo trascorso assieme e per sapermi ascoltare. Eric, thank you for your plain speaking and caustic sense of humor. Christian, thanks for all the beer, for the endless discussions about movies and for making fun of stupidity. Giuseppe 'cervello in fuga', with whom I have consumed an insane amount of cappuccinos at 'il maltese' while discussing about Italian politics. Martin for being a great sports partner and an optimistic billiards player.

Last, but not least I thank my dear family – especially my beloved mother Lucia for her endless love.

Nils-Ole Walliser
Stuttgart, Germany
October 2012

Appendix A

Appendix to chapter 4

A.1 Definitions and rules for B-branes

A.1.1 D-brane charges

We will be studying D-branes wrapped on even-dimensional submanifolds of a CY threefold X in type IIB theory, or *B-branes*. More precisely, we will deal with spacetime filling D7- and D3-branes on one hand, and Euclidean D3-branes, henceforth referred to as E3-branes, that are point-like in four dimensions.

For a D-brane wrapped on a submanifold P, the coupling to the total R-R-potential $C = C_0 + C_2 + C_4 + C_6 + C_8$ is given by the following:

$$S^{\text{Dbrane}}_{P,C} = 2\pi \int_P C \wedge e^{-B} \operatorname{Tr} e^F \sqrt{\frac{\widehat{A}(TP)}{\widehat{A}(N_P)}}, \qquad (A.1)$$

where B is the NS-NS two-form (pulled-back onto P), F is the quantized $U(1)$ field-strength of the world-volume theory, $\widehat{A}$ is the 'A-roof genus', and TP and N_P are the tangent and normal bundles of P, respectively. Define a polyform $\Gamma \in H^{\text{even}}(X, \mathbb{Z})$ such that

$$S^{\text{Dbrane}}_{P,C} = 2\pi \int_P C \wedge e^{-B} \Gamma. \qquad (A.2)$$

Then Γ is interpreted as a source for R-R-charges, or a charge vector.

Throughout this work, we will neglect possible torsion charges and will always deal with divisors with $b^3 = 0$, such that the pullback of the NS-NS three-form H is trivial on our D-branes. Hence, we can define D-brane charge by means of the cohomology of the internal space X. The most general 'charge vector' Γ will be of the form

$$\Gamma = q_{D9} + q^A_{D7} D_A + q_{D5,A} \tilde{D}^A + q_{D3}\, \omega, \qquad (A.3)$$

where the D_A and $\tilde{D}^A$ define a base for $H^2(X, \mathbb{Z})$ and $H^4(X, \mathbb{Z})$, respectively, and ω is the volume-form of X.

For instance, for a single *smooth* D7-brane wrapped on a four-dimensional submanifold, i.e. a divisor P, with inclusion map

$$\imath : P \hookrightarrow X, \qquad (A.4)$$

that is Poincaré dual to a two-form $[D_P] \in H^2(X, \mathbb{Z})$, supporting a $U(1)$ field-strength F, the total charge vector will be given by the following:

$$\Gamma_{D7} = [D_P] + \left(\int_P F \cdot \imath^*(D_A)\right) \tilde{D}^A + \left(\frac{1}{2} \int_P F^2 + \frac{\chi(P)}{24}\right) \omega. \tag{A.5}$$

In this notation, the large volume formula for the FI-term induced by a D7-brane is very simple:

$$\xi = \text{Im}\left(-\frac{1}{\mathcal{V}} \int_X e^{-(B+iJ)} \Gamma_{D7}\right) = \frac{1}{\mathcal{V}} \int_{D7} J \cdot (F - B), \tag{A.6}$$

where the F contains the half-integral flux.

For two stacks of (magnetized) branes D_1 and D_2, the net number of chiral bifundamental strings stretching from D_1 to D_2 is given by the so-called Dirac-Schwinger-Zwanziger intersection product:

$$\langle \Gamma_1, \Gamma_2 \rangle \equiv \int_X \Gamma_1 \wedge \Gamma_2^*, \tag{A.7}$$

where Γ^* is defined by flipping the sign of the two- and six-form components

$$\Gamma^* = q_{D9} - q_{D7}^A D_A + q_{D5,A} \tilde{D}^A - q_{D3} \omega. \tag{A.8}$$

In the case of two intersecting D7-branes with $U(1)$ fluxes F_1 and F_2, respectively, this reduces to

$$\int_{D7_1 \cap D7_2} (F_2 - F_1). \tag{A.9}$$

Note that, by construction, the Γ charge vectors project out any flux on a D7-brane that is trivial in the CY even though it may be non-trivial on the divisor. Since the DSZ product depends only on the Γ's, this means that it gives information about the net chiral spectrum, but misses possible vector-like pairs. Given two D7-branes intersecting at a Riemann surface, it is possible to count such non-chiral fields by counting sections of the corresponding bifundamental bundle over the surface. However, this is beyond the scope of this work.

A.1.2 Orientifolding

In this work, we will consider holomorphic involutions of the form $x_i \longrightarrow -x_i$, where x_i is a homogeneous coordinate. Involutions of this type act trivially on the (co)homology of the threefold, which implies that $H^2_-(X, \mathbb{Z}) = 0$, i.e. only invariant two-forms exist. We will look for involutions of O7/O3-type.

The action on the massless closed string fields is [166]

$$\begin{aligned} C_{0,4,8} &\to \sigma^* C_{0,4,8}, \quad C_{2,6,10} \to -\sigma^* C_{2,6,10}, \quad g \to \sigma^* g, \\ B &\to -\sigma^* B \mod H^2(X, \mathbb{Z}). \end{aligned} \tag{A.10}$$

where the B-field is only well-defined up to an integral cohomology shift. This implies that D9- and D5-charges flip sign under this action, whereas D7- and D3-charges remain intact. The

D7 on O7	transversally inv. D7	D7/image-D7 pair	transversally inv. E3
$O(n)$	$Sp(2\,n)$	$U(n)$	$O(n)$

Table A.1: Configurations of D7's and O7, and their associated gauge groups.

action on a world-volume gauge field living on an orientifold-invariant D3- or D7-brane stack is [167]

$$A \to -M\sigma^* A^t M^{-1}, \tag{A.11}$$

where M is a symmetric or antisymmetric matrix depending on the gauge group surviving on the D-brane. We will choose the world-sheet orientifold projection such that D7-branes wrapped on a O7-plane have $O(n)$ gauge groups. For a general brane, the gauge group is decided by counting the number DN directions with respect to a reference D7-brane on the O7-plane. For instance, a transversally invariant D7-brane, i.e. a brane that is mapped to itself but does not lie on top of the O7-plane, will have a symplectic gauge group. On the other hand, a pair of D7-branes that are exchanged by the involution will have a unitary group, since they are not affected by the orientifolding. A transversally invariant Euclidean D3-brane, or E3-brane, will have an orthogonal group. This is summarized in table A.1:

Transversally invariant D7-branes necessarily satisfy the restriction that they always intersect the O7-plane at double points. As explained in [72], such branes are wrapped on divisors given by equations of the form $\eta^2 + \xi^2 \chi = 0$, where η and χ are generic polynomials of appropriate degree, and $\xi = 0$ is the locus of the O7-plane. This mimics the equation of the *Whitney umbrella*. When we use such branes, we will refer to them as *Whitney-type* branes for simplicity.

For open world-sheets, the B-field and $U(1)$ field-strength F transform as follows:

$$B \to -\sigma^*(B) + \Lambda, \quad F \to -\sigma^*(F) + \imath^*(\Lambda), \tag{A.12}$$

where $\Lambda \in H^2(X, \mathbb{Z})$, such that the gauge invariant combination $\mathcal{F} \equiv F - \imath^*(B)$ transforms as

$$\mathcal{F} \to -\sigma^*(\mathcal{F}). \tag{A.13}$$

We will use the following gauge $\Lambda = 2\,B$, such that

$$B \to +\sigma^*(B) \quad \text{and} \quad F \to -\sigma^*(F) + 2\,\sigma^*(B). \tag{A.14}$$

The involutions considered in chapter 4 act trivially on the even cohomology of X, i.e. $h^{1,1}_- = 0$. Hence, the 'σ^*' can be dropped.

Given a fixed point locus of the involution consisting of O7- and O3-planes, the total charge vector will be:

$$\Gamma = \Gamma_{O7} + \Gamma_{O3} = [D_{O7}] + \frac{\chi(O7)}{24}\omega + N_{O3}. \tag{A.15}$$

A.1.3 K-theory construction of D7-branes

In this appendix, we concisely explain how to describe D7-branes using the picture developed in [72]. There are two types of D7-branes we wish to describe: D7/image-D7 pairs, and transversally invariant, *Whitney-type* branes.

Brane/image-brane pairs

Let us begin by the former. Suppose we want to write the charge vector Γ_{D7} of a D7-brane wrapped on the divisor D_P, and its orientifold image Γ'_{D7}, which is wrapped on a divisor in the same homology class.[1] We introduce two D9/$\overline{D9}$ pairs with fluxes as follows

$$D9_1 : F_1 = D_P - S, \qquad \overline{D9}_1 : F'_1 = S - D_P + 2B,$$
$$D9_2 : F_2 = S + 2B, \qquad \overline{D9}_2 : F'_2 = -S, \qquad (A.16)$$

where $S \in H^2(X, \mathbb{Z})$. The respective charge vectors $\Gamma_1, \Gamma'_1, \Gamma_2, \Gamma'_2$ are expressed as follows:

$$\Gamma = \operatorname{ch}(F) \sqrt{\operatorname{td}(X)} = \operatorname{ch}(F)\left(1 + \frac{c_2(X)}{24}\right). \qquad (A.17)$$

Now, we can simply write

$$\Gamma_{D7} = \Gamma_1 + \Gamma'_2 = \left(\operatorname{ch}(F_1) - \operatorname{ch}(F'_2)\right)\left(1 + \frac{c_2(X)}{24}\right). \qquad (A.18)$$

It is easy to see that this expression has vanishing D9-charge, and that its D7-charge is equal to $F_1 - F'_2 = D_P$, as desired. After tachyon condensation, the shift two-form S translates into a flux on the D7-brane equal to $F_{D7} = \frac{1}{2} D_P - S$. One can easily check that this charge vector indeed reproduces the right flux and curvature induced lower brane charges expected for a D7-brane (A.5). The charge vector of the image-brane is defined by using the image D9/$\overline{D9}$ pair as $\Gamma_{D7'} = \Gamma'_1 + \Gamma_2$. One can easily check the the D7-charge will be D_P again, and that the flux on the resulting D7 will be $-F_{D7} + 2B$.

For the sake of concreteness, let us work out the charge vector of the $D7_A$-brane of the first scenario of the first model in section 4.4. In this case, the divisor and the desired world-volume flux are

$$D_P = \eta_4, \quad F_A = \tfrac{1}{2}\eta_4 - S, \quad \text{with} \quad S = -\sum_{i=1}^{4} a_i \eta_i.$$

We can immediately write the charge vector as

$$\begin{aligned}\Gamma &= \left(\exp(\eta_4 - S) - \exp(-S)\right)\left(1 + \frac{c_2(X)}{24}\right) \\ &= \eta_4 + \eta_4 \cdot (\tfrac{1}{2}\eta_4 - S) + \left(\tfrac{1}{2}\eta_4(\tfrac{1}{2}\eta_4 - S)^2 + \eta_4^3 + c_2(X)\eta_4\right),\end{aligned} \qquad (A.19)$$

where the last two terms give the Euler characteristic of the divisor, $\eta_4^3 + c_2(X) \cdot \eta_4 = \chi(\eta_4)$. Hence, we see that it is clearly much more convenient to construct charge vectors by means of this method than by straightforward application of (A.5). As explained at the end of A.1.1, this charge vector can only be used to compute induced charges, and deduce the net chiral spectrum of intersecting D7-branes. To find out about the non-chiral sector, more work is required.

---

[1] We consider only involutions such that $h^{1,1}_-(X) = 0$.

Whitney-type branes

Let us now review how Whitney-type branes are treated in this language. In [72], the condition was derived that all orientifold-invariant configurations should actually be made out of an even number of D9/$\overline{\text{D9}}$ pairs. In a sense, a Whitney-type brane can be thought of as a D7/image-D7 pair that has recombined into a single invariant brane. For a Whitney-type brane of even D7-charge $D_W = 2 D_P$, the charge vector is simply:

$$\begin{aligned}\Gamma &= \Gamma_1 + \Gamma_2 + \Gamma_1' + \Gamma_2' \\ &= \Big(\text{ch}(D_P - S) + \text{ch}(S + 2B) - \text{ch}(-D_P + S + 2B) - \text{ch}(-S)\Big)\left(1 + \frac{c_2(X)}{24}\right),\end{aligned} \qquad (A.20)$$

where the Γ's are the ones we defined before. One can easily check that this is involution-invariant, that the two-form component is indeed D_W, and that the four-form component is $D_W \cdot F_W = D_W \cdot B$, as expected. The choice of the two-form S corresponds to adjusting the flux on the D7$_W$ of type $h_-^{1,1}(D_H)$. Define the involution as $\xi \to -\xi$. Then the D7$_W$-brane resulting from tachyon condensation will have a singular divisor equation given by

$$\eta^2 + \xi^2 \left(\rho \tau - \psi^2\right) = 0, \qquad (A.21)$$

where $\{\eta; \psi; \rho; \tau\}$ are sections of the line bundles associated to the classes

$$\{D_P \,;\, D_P - D_\xi \,;\, 2\left(D_P - S - B\right) - D_\xi \,;\, 2\left(S + B\right) - D_\xi\}, \qquad (A.22)$$

respectively. In order for the D7-brane to retain its 'structural integrity', one must choose S such that all these bundles are positive definite, or else this will modify the structure of the brane severely. Starting with ψ, we see that as long as we do not choose to have a single D7/image-D7 pair on top of the O7-plane, this class will always be positive-definite. Should either one of the polynomials ρ and τ correspond to a section of a negative bundle, which would not be globally well-defined, then we would have to set it identically to zero. In this case, the divisor equation would factorize into a D7/image-D7 pair as follows:

$$\eta^2 + \xi^2 \psi^2 = 0 \quad \Longrightarrow \quad \{\eta + \xi\psi = 0\} \cup \{\eta - \xi\psi = 0\}. \qquad (A.23)$$

The constraints for ρ and τ to be globally well-defined are

$$D_P - \frac{[\xi]}{2} - B \geq S \geq \frac{[\xi]}{2} - B. \qquad (A.24)$$

Fortunately, S will drop out of the calculation of intersection products with the other present branes. It will, however, enter the D3 tadpole calculation.

A.2 Third model

A.2.1 R2 resolution of the $\mathbb{P}^4_{2,1,6,1,2}(12)/\mathbb{Z}_2 : 0\,0\,1\,1\,0$ geometry

We repeat the projective weights for this space for convenience in table A.2. The Stanley-

	x_1	x_2	x_3	x_4	x_5	x_6	x_7	x_8	p
	2	1	6	1	2	0	0	0	12
	2	1	5	0	2	0	0	2	12
	2	0	5	1	2	0	2	0	12
	1	0	3	0	1	1	0	0	6

Table A.2: Projective weights for the R1 resolution of $\mathbb{P}^4_{2,1,6,1,2}(12)/\mathbb{Z}_2 : 0\,0\,1\,1\,0$.

Reisner ideal reads

$$I_{SR} = \{x_2x_3,\, x_2x_4,\, x_3x_4,\, x_3x_6,\, x_2x_8,\, x_6x_8,\, x_1x_4x_5x_7,\, x_1x_5x_6x_7,\, x_1x_5x_7x_8\}. \tag{A.25}$$

The triple intersection numbers in the basis $\eta_1 = D_2$, $\eta_2 = D_4$, $\eta_3 = D_6$ and $\eta_4 = D_8$ are encoded in

$$\begin{aligned}I_3 &= 9\eta_1^3 + 3\eta_2^3 + 8\eta_3^3 - 72\eta_4^3 - 3\eta_1^2\eta_3 \\ &\quad + 3\eta_2^2\eta_3 - 12\eta_2^2\eta_4 + \eta_1\eta_3^2 - 5\eta_2\eta_3^2 + 30\eta_2\eta_4^2\,.\end{aligned} \tag{A.26}$$

The volumes of the corresponding divisors are

$$\begin{aligned}\tau_1 &= \frac{1}{2}(3t_1 - t_3)^2\,, \\ \tau_2 &= \frac{1}{2}\left(3t_2^2 + 6t_2t_3 - 5t_3^2 - 24t_2t_4 + 30t_4^2\right), \\ \tau_3 &= -\frac{1}{2}(t_1 + t_2 - 2t_3)(3t_1 - 3t_2 + 4t_3)\,, \\ \tau_4 &= -6(t_2 - 3t_4)(t_2 - 2t_4)\,.\end{aligned} \tag{A.27}$$

The volume of the CY manifold is given by

$$\begin{aligned}\mathcal{V} &= \frac{1}{6}\left[-\frac{1}{15}(3t_2 - 5t_3)^3 - \frac{1}{3}(-3t_1 + t_3)^3 - \frac{3}{5}(5t_4 - 2t_2)^3 + 3t_4^3\right] \\ &= \frac{\sqrt{2}}{3}\left[\frac{1}{2\sqrt{6}}(\tau_1 + 5\tau_2 + 3\tau_3 + 2\tau_4)^{\frac{3}{2}} - \frac{1}{10\sqrt{6}}(\tau_1 + 5\tau_2 + 3\tau_3)^{\frac{3}{2}} - \frac{1}{15}(\tau_1 + 3\tau_3)^{\frac{3}{2}} - \frac{1}{3}\tau_1^{\frac{3}{2}}\right].\end{aligned} \tag{A.28}$$

It has the expected Swiss cheese form. From this volume formula we deduce the diagonal basis to be

$$\begin{aligned}D_a &= \eta_1 + 5\eta_2 + 3\eta_3 + 2\eta_4\,, \\ D_b &= \eta_1 + 5\eta_2 + 3\eta_3\,, \\ D_c &= \eta_1 + 3\eta_3\,, \\ D_d &= \eta_1\,.\end{aligned} \tag{A.29}$$

In this basis the total volume reads

$$\mathcal{V} = \frac{\sqrt{2}}{3}\left(\frac{1}{2\sqrt{6}}\tau_a^{\frac{3}{2}} - \frac{1}{10\sqrt{6}}\tau_b^{\frac{3}{2}} - \frac{1}{15}\tau_c^{\frac{3}{2}} - \frac{1}{3}\tau_d^{\frac{3}{2}}\right), \tag{A.30}$$

and the triple intersections can be rewritten as

$$I_3 = 24\,D_a^3 + 600\,D_b^3 + 225\,D_c^3 + 9\,D_d^3\,. \tag{A.31}$$

The Kähler cone conditions read as follows:

$$\begin{aligned} t_2 - 2t_4 &> 0\,,\\ -t_2 + t_3 + t_4 &> 0\,,\\ t_1 + t_2 - 2t_3 &> 0\,,\\ -3t_1 + t_3 &> 0\,. \end{aligned} \tag{A.32}$$

Searching for smooth, 'small' cycles with holomorphic Euler characteristic equal to one, we find the following three surfaces:

$$\{D_2, D_4, D_6\} = \{\eta_1, \eta_2, \eta_3\}\,, \quad \text{with} \quad h^{1,1} = \{1,7,2\}\,. \tag{A.33}$$

The first surface is a $\mathbb{CP}^2$. By inspecting the intersection numbers, we see that the other two surfaces fail to be del Pezzo surfaces, even though their Hodge numbers are consistent with those of dP_6, and the Hirzebruch surfaces $\mathbb{F}_n$ (for arbitrary n), respectively. Let us work out the topology of D_4 in more detail.[2] The SR ideal shows that x_2 and x_3 can not vanish on the surface. Hence, we gauge-fix them to one. Now the CY polynomial takes the following form:

$$P^{(6)}(x_1; x_5; x_6\,x_7) + x_7\,x_8 = 0\,, \tag{A.34}$$

where the first term is some polynomial of degree six in the three arguments given. We can now define a map from this surface onto $\mathbb{CP}^2$ as follows:

$$(x_1 : x_5 : x_6 : x_7 : x_8) \mapsto (y_1 : y_2 : y_3) = (x_1 : x_5 : x_6\,x_7)\,. \tag{A.35}$$

This map is a blow-down of our surface onto the projective plane. Now, we can distinguish two cases:

1. $x_7 \neq 0$. In this case, we gauge-fix $x_7 = 1$. Now we see that choosing a point on the $\mathbb{CP}^2$, which amounts to choosing x_1, x_2, and x_6, completely determines x_8, since it appears linearly in the CY equation.

2. $x_7 = 0$. In this case, the CY equation takes the form $P^{(6)}(x_1; x_5) = 0$, and x_6 and x_8 are undetermined. This means, that the preimages of the six points on the $\mathbb{CP}^2$ with $(x_1 : x_5 : 0)$, where the $P^{(6)}(x_1; x_5) = 0$ is satisfied are curves parametrized by $(x_6 : x_8)$.

Therefore, our surface is indeed the blow-up of the projective plane at six points. However, all six points lie on the line (the $\mathbb{CP}^1$) defined by $y_3 = 0$. Hence, they are not in generic positions, which is a requirement in order to have a dP_6.

Scenario	E3	D7$_A$	D7$_B$
I	η_1	η_2	η_3
	arbitrary	arbitrary	$\{1+e_1+n\,;b_2\,;1+e_3+3\,n\,;b_4\}$
II	η_2	η_1	η_3
	arbitrary	arbitrary	$\{b_1\,;3+e_2+5\,n\,;1+e_3+3\,n\,;b_4\}$

Table A.3: Two 'local' models.

A.2.2 Scenarios in the third model

By inspecting the intersection numbers of this CY we see that η_1 and η_2 do not intersect. Therefore, we only have two possible scenarios. We summarize our results in table A.3.

For the global model, we will pick the involution $x_3 \to -x_3$. Solving the equations

$$\langle \Gamma_W, \Gamma_A \rangle = \langle \Gamma_W, \Gamma_B \rangle = 0, \tag{A.36}$$

we find the following solutions:

1. **Scenario I**: The constraints we get from setting the chiral intersections with the hidden brane to zero are the following:

$$N_A = 3\,N, \qquad N_B = 5\,N, \qquad \text{for some} \quad N \in \mathbb{Z}, \tag{A.37}$$
$$a_2 = 2 + e_2 + 5\,t, \qquad a_4 = 1 + e_4 + 2\,t, \qquad \text{for some} \quad t \in \mathbb{Z}. \tag{A.38}$$

As we see here, this setup requires that we put further constraints on the 'local' model. To get an idea of how much D3-tadpole this Whitney-type brane can induce, let us compute it for the 'minimal' choice of the shift vector S in formula (A.24):

$$Q_{W,D3} = 372 - \frac{3}{2}N - 197\,N^3. \tag{A.39}$$

Finally, let us compute the FI-terms for both MSSM branes in light of these constraints:

$$\xi_A, \xi_B \propto \sqrt{\tau_c}, \tag{A.40}$$

where $D_c = \eta_1 + 3\,\eta_3$.

The self-intersection volume of the instanton in this scenario is given by

$$\text{Vol}\,(D_{E3} \cap D_{E3}) = 9t_1 - 3t_3 = -3\sqrt{2}\sqrt{\tau_b}. \tag{A.41}$$

Looking at the Kähler cone in the diagonal basis

$$\begin{aligned} \sqrt{\tau_a} - 5\sqrt{\tau_b} &> 0, \\ 2\sqrt{\tau_b} - 3\sqrt{\tau_c} &> 0, \\ 5\sqrt{\tau_c} - \sqrt{\tau_d} &> 0, \\ 3\sqrt{\tau_d} &> 0. \end{aligned} \tag{A.42}$$

---
[2]We are very grateful to H. Skarke for sharing this calculation with us.

x_1	x_2	x_3	x_4	x_5	x_6	x_7	x_8	p
1	1	3	1	3	0	0	0	9
2	2	3	2	0	0	0	9	18
0	0	0	0	1	2	3	0	6
0	0	0	0	0	1	1	1	3

Table A.4: Projective weights for the resolved $\mathbb{P}^4_{1,1,3,1,3}(9)/\mathbb{Z}_3 : 0\,0\,2\,1\,0$ space.

The third equation indicates that the volume of the instanton has to go to zero. The reason is that the D-term potential dominates in the LVS, and setting this term to zero means τ_c has to vanish. Having a volumeless instanton now ruins the LVS. Again, we expect this D-term to be corrected by string loops, which could salvage this LVS.

2. **Scenario** II: the only constraint we get from setting the chiral intersections to zero is $N_B = 3\,N_A$. Let us also compute the D3-tadpole for this hidden brane with the 'minimal' choice of S:

$$Q_{W,D3} = 372 + \frac{3}{2} N_A - 75\, N_A^3 \,. \tag{A.43}$$

In this case, both branes give again similar FI-terms:

$$\xi_A,\, \xi_B \propto \sqrt{\tau_1} \,. \tag{A.44}$$

The self-intersection volume of the instanton in this scenario is given by

$$\begin{aligned}\mathrm{Vol}\left(D_{E3}\cap D_{E3}\right) &= 3t_2 + 3t_3 - 12t_4 = -4\sqrt{3\tau_a} + \frac{1}{2}\sqrt{3\tau_b} - \frac{21}{2\sqrt{2}}\sqrt{\tau_c} \\ &= -\left(4\sqrt{3\tau_a} - \frac{1}{2}\sqrt{3(\tau_c + 5\tau_{E3})} + \frac{21}{2\sqrt{2}}\sqrt{\tau_c}\right)\,.\end{aligned} \tag{A.45}$$

In this case, the same problem as in (4.76) occurs, by making the volume large we get an imaginary part in the solution for the volume. Thus, in the second scenario we do not get a large volume compactification either.

A.3   Fourth model: a matterless model

The following model, as it turns out, will yield a trivial field content. Nevertheless, we will present the geometry in case the reader wants to use it differently.

A.3.1   The resolved $\mathbb{P}^4_{1,1,3,1,3}(9)/\mathbb{Z}_3 : 0\,0\,2\,1\,0$ geometry

The Stanley-Reisner ideal reads

$$I_{SR} = \{x_3 x_5,\ x_5 x_7,\ x_6 x_7,\ x_3 x_8,\ x_6 x_8,\ x_1 x_2 x_4\}\,. \tag{A.46}$$

The triple intersection numbers in the basis $\eta_1 = D_8$, $\eta_2 = D_6$, $\eta_3 = D_5$, $\eta_4 = D_{1,2,4}$ are encoded in

$$\begin{aligned} I_3 &= -216\eta_1^3 + 9\eta_2^3 + 9\eta_3^3 + \eta_3\eta_4^2 + \eta_2\eta_4^2 - 3\eta_3^2\eta_4 \\ &\quad -27\eta_1\eta_3^2 - 3\eta_2^2\eta_4 - 18\eta_1^2\eta_4 + 81\eta_1^2\eta_3 + 9\eta_1\eta_3\eta_4 \,. \end{aligned} \qquad (A.47)$$

The volumes of the corresponding divisors are

$$\begin{aligned} \tau_1 &= -\frac{9}{2}(2t_1 - t_3)(12t_1 - 3t_3 + 2t_4)\,, \\ \tau_2 &= \frac{1}{2}(3t_2 - t_4)^2\,, \\ \tau_3 &= \frac{1}{2}(9t_1 - 3t_3 + t_4)^2\,, \\ \tau_4 &= \frac{1}{2}\left(-18t_1^2 - 3t_2^2 + 18t_1 t_3 - 3t_3^2 + 2t_2 t_4 + 2t_3 t_4\right)\,. \end{aligned} \qquad (A.48)$$

The volume of the Calabi-Yau manifold is given by

$$\begin{aligned} \mathcal{V} &= \frac{1}{18}\left[3(3t_1 + t_4)^3 - t_4^3 - (3t_2 - t_4)^3 - (9t_1 - 3t_3 + t_4)^3\right] \\ &= \frac{\sqrt{2}}{9}\left[\frac{1}{\sqrt{3}}(\tau_1 + 3\tau_3)^{\frac{3}{2}} - (\tau_1 - \tau_2 + 2\tau_3 - 3\tau_4)^{\frac{3}{2}} - \tau_2^{\frac{3}{2}} - \tau_3^{\frac{3}{2}}\right]\,. \end{aligned} \qquad (A.49)$$

It has the expected Swiss cheese form. From this volume formula we deduce the diagonal basis to be

$$\begin{aligned} D_a &= \eta_1 + 3\eta_3\,, \\ D_b &= \eta_1 - \eta_2 + 2\eta_3 - 3\eta_4\,, \\ D_c &= \eta_2\,, \\ D_d &= \eta_3\,, \end{aligned} \qquad (A.50)$$

in this basis the total volume reads

$$\mathcal{V} = \frac{\sqrt{2}}{9}\left[\frac{1}{\sqrt{3}}\tau_a^{\frac{3}{2}} - \tau_b^{\frac{3}{2}} - \tau_c^{\frac{3}{2}} - \tau_d^{\frac{3}{2}}\right]\,, \qquad (A.51)$$

and the triple intersections can be rewritten as

$$I_3 = 27 D_a^3 + 9 D_b^3 + 9 D_c^3 + 9 D_d^3\,. \qquad (A.52)$$

The Kähler cone conditions read as follows:

$$\begin{aligned} -2t_1 + t_3 &> 0\,, \\ t_1 &> 0\,, \\ 3t_1 + t_2 - t_3 &> 0\,, \\ -3t_2 + t_4 &> 0\,, \\ t_2 &> 0\,. \end{aligned} \qquad (A.53)$$

Searching for rigid divisors with holomorphic Euler characteristic one, we find the following three solutions:

$$\{D_3, D_5, D_6\} = \{-3\,\eta_4 + 2\,\eta_3 - \eta_2 + \eta_1\,, \eta_3\,, \eta_2\}\,, \quad \text{with} \quad h^{1,1} = \{1,1,1\}\,. \tag{A.54}$$

Hence, all three are $\mathbb{CP}^2$'s. The striking feature about these divisors, which ultimately kills the model for our purposes, lies in the fact that no two of them intersect. Although this automatically solves the problem of unwanted zero modes, it does so too drastically, as no chiral matter can arise from D7-branes wrapped on them.

Inspecting the available involutions, we see that it is impossible to have two D7-branes on distinct cycles and cancel the D7-tadpole. Hence, one can only have a single D7-brane, and in this case, it must be on top of the O7-plane.

A.3.2 Moduli stabilization

Although we can not do any model building in this example we can nevertheless look at the stabilization problem. So first we choose divisors on which we would like to wrap our D7-branes. However, by inspecting table A.3.1 carefully, we see that we can have only one brane if we want to compensate D7-charge only via an orientifold plane. Hence, we will work with a configuration where we have a D7-brane on a divisor four times the divisor class of the orientifold and nothing else. If we wrap the brane on a diagonal divisor, we obtain that the FI-term is proportional to its volume. Knowing this we have to take a divisor that is unrestrictedly shrinkable. For this we rewrite the Kähler cone in terms of the diagonal basis

$$\begin{aligned}
\sqrt{\tau_a} - \sqrt{\tau_c} &> 0\,, \\
\sqrt{\tau_a} - \sqrt{\tau_b} &> 0\,, \\
\sqrt{\tau_c} - \sqrt{\tau_d} &> 0\,, \\
3\sqrt{\tau_d} &> 0\,, \\
\sqrt{\tau_b} - \sqrt{\tau_d} &> 0\,.
\end{aligned} \tag{A.55}$$

Having the D7-brane on D_d we can put the instanton either on D_b or D_c. For these two cases, we get the following self-intersection volumes for the instanton:

$$\text{Vol}\,(D_b \cap D_b) = -3t_4 = -3\sqrt{2\tau_b}\,, \tag{A.56}$$

and

$$\text{Vol}\,(D_c \cap D_c) = -3t_4 + 9t_1 - 27t_3 = -3\sqrt{2\tau_c}\,, \tag{A.57}$$

respectively. In this case, the potentials for the two scenarios are symmetric under exchange of τ_b and τ_c. With (4.12) and $A_{E3} = 1$, $|W_0| = 5$ and $g_s = \frac{1}{10}$ we find for both minima

$$\tau_{E3} = 1.41\,, \quad \mathcal{V} = 6.74 \cdot 10^{36}\,, \tag{A.58}$$

where in each case one flat direction remains. Therefore, in this trivial model we can only stabilize three out of the four moduli.

Appendix B

Appendix to chapter 5

B.1 Matter genera and Yukawa points

In the following table, we list the matter genera and Yukawa numbers for those del Pezzos, where the F-theory GUT lives on a Calabi-Yau fourfold described by to a reflexive polytope, where at least one nef partition is compatible with the elliptic fibration. Furthermore, the base B should be regular, and at least one of the del Pezzos inside the base should admit a decoupling limit. Note that for the calculation of these numbers the formulas (5.20) and (5.21) have been used. There it has been assumed that the curves involved are irreducible. Since we could not check this explicitly for every model, some of these numbers might be incorrect.

		$SU(5)$			$SO(10)$				
type	$g_{SU(6)}$	$g_{SO(10)}$	n_{E_6}	$n_{SO(12)}$	$g_{SO(12)}$	g_{E_6}	n_{E_7}	$n_{SO(14)}$	#
dP_8	11	1	0	0	2	1	2	12	9
dP_7	57	1	4	6	7	3	12	48	187
	102	2	10	14	12	6	22	76	2
	75	1	6	9	9	4	16	60	5
	21	1	0	0	3	1	4	24	73
	48	0	2	4	6	2	10	44	1
	66	0	4	7	8	3	14	56	2
dP_6	85	1	6	9	10	4	18	72	161
	31	1	0	0	4	1	6	36	47
	58	0	2	4	7	2	12	56	32
	130	2	12	17	15	7	28	100	3
	76	0	4	7	9	3	16	68	4
	103	1	8	12	12	5	22	84	3
dP_5	68	0	2	4	8	2	14	68	96
	113	1	8	12	13	5	24	96	340
	104	0	6	10	12	4	22	92	7
	131	1	10	15	15	6	28	108	14

	158	2	14	20	18	8	34	124	17
	86	0	4	7	10	3	18	80	34
	41	1	0	0	5	1	8	48	47
	185	3	18	25	21	10	40	140	3
	176	2	16	23	20	9	38	136	1
dP_4	141	1	10	15	16	6	30	120	141
	96	0	4	7	11	3	20	92	56
	78	0	2	4	9	2	16	80	60
	186	2	16	23	21	9	40	148	16
	51	1	0	0	6	1	10	60	21
	114	0	6	10	13	4	24	104	23
	159	1	12	18	18	7	34	132	4
	132	0	8	13	15	5	28	116	10
dP_3	124	0	6	10	14	4	26	116	189
	169	1	12	18	19	7	36	144	267
	205	1	16	24	23	9	44	169	28
	160	0	10	16	18	6	34	140	6
	268	4	26	36	30	14	58	204	10
	214	2	28	26	24	10	46	172	18
	88	0	2	4	10	2	28	92	63
	61	1	0	0	71	1	12	72	32
	142	0	8	13	16	5	30	128	45
	106	0	4	7	12	3	22	104	35
	187	1	14	21	21	8	40	156	15
	250	2	22	32	28	12	54	196	1
	241	3	22	31	27	12	52	188	5
dP_2	170	0	10	16	19	6	36	152	218
	134	0	6	10	15	4	28	128	180
	197	1	14	21	22	8	42	168	427
	215	1	16	24	24	9	46	180	102
	269	3	24	34	30	13	58	212	25
	116	7	4	7	13	3	24	116	73
	242	2	20	29	27	11	52	196	105
	188	0	12	19	21	7	40	164	18
	71	1	0	0	8	1	14	84	30
	152	0	8	13	17	5	32	140	117
	260	2	22	32	29	12	56	208	22
	323	5	32	44	36	17	70	244	10
	98	0	2	4	11	2	20	104	34
	296	4	28	39	33	15	64	228	19

	206	0	14	22	23	8	44	176	1
	305	3	28	40	34	15	66	236	2
dP_1	225	1	16	24	25	9	48	192	1150
	252	0	18	28	28	10	54	212	11
	144	0	6	10	16	4	30	140	482
	81	1	0	0	9	1	16	96	214
	198	0	12	19	22	7	42	176	139
	270	2	22	32	30	12	58	220	239
	180	0	10	16	20	6	38	164	603
	162	0	8	13	18	5	34	152	476
	315	3	28	40	35	15	68	248	54
	378	6	38	52	42	20	82	284	20
	108	0	2	4	12	2	22	116	278
	441	9	48	64	49	25	96	320	9
	234	0	16	25	26	9	50	200	7
	243	1	18	27	27	10	52	204	51
	216	0	14	22	24	8	46	188	28
	297	3	26	37	33	14	64	236	54
	324	4	30	42	36	16	70	252	27
	126	0	4	7	14	3	26	128	175
	351	5	34	47	39	18	76	268	15
	270	0	20	31	30	11	58	224	1
dP_0	253	1	18	27	28	10	54	216	338
	496	10	54	72	55	28	108	360	12
	91	1	0	0	10	1	18	108	150
	190	0	10	16	21	6	40	176	763
	325	3	28	40	36	15	70	260	126
	136	0	4	7	15	3	28	140	380
	406	6	40	55	45	21	88	308	33

Table B.1: Topological numbers of del Pezzos with physical decoupling limit.

Appendix C

Appendix to chapter 6

C.1 Expansions around LCS points

C.1.1 One-parameter models

For a one-parameter model, the period vector takes the following general form around the LCS point [141]

$$\begin{pmatrix} \Pi_3 \\ \Pi_2 \\ \Pi_1 \\ \Pi_0 \end{pmatrix} \sim \begin{pmatrix} \alpha_3 t^3 + \gamma_3 t + i\delta_3 \\ \beta_2 t^2 + \gamma_2 t + \delta_2 \\ t \\ 1 \end{pmatrix}. \tag{C.1}$$

Here $t \sim -i \log z$, and the LCS point is at $\operatorname{Im} t \to \infty$. All coefficients except δ_3 are rational. For the models we study, the coefficients are presented in table C.1.

Let $t = t_1 + it_2$ with $t_{1,2} \in \mathbb{R}$. For general expansion coefficients we then get the following expansions around $t_2 = \infty$

$$e^{-K} = (-2\beta_2 - 2\alpha_3)t_2^3 + (2\delta_2 + 6\alpha_3 t_1^2 - 2\beta_2 t_1^2 + 2\gamma_3)t_2 + \dots, \tag{C.2}$$

$$\mathcal{G}_t = \begin{pmatrix} g_{11} t_2^3 + \mathcal{O}(t_2) & g_{12} t_2 + \mathcal{O}(1/t_2) & g_{13} t_2 + \mathcal{O}(1/t_2) & g_{14}\frac{1}{t_2} + \mathcal{O}(1/t_2^3) \\ \cdot & g_{22} t_2 + \mathcal{O}(1/t_2) & g_{23}\frac{1}{t_2} + \mathcal{O}(1/t_2^3) & g_{24}\frac{1}{t_2} + \mathcal{O}(1/t_2^3) \\ \cdot & \cdot & g_{33}\frac{1}{t_2} + \mathcal{O}(1/t_2^3) & g_{34}\frac{1}{t_2^3} + \mathcal{O}(1/t_2^5) \\ \cdot & \cdot & \cdot & g_{44}\frac{1}{t_2^3} + \mathcal{O}(1/t_2^5) \end{pmatrix} \cdot \tag{C.3}$$

The coefficients g_{ij} are a little messy:

$$g_{11} = -\frac{2\alpha_3^2 (9\alpha_3 + 5\beta_2)}{(\alpha_3 + \beta_2)(9\alpha_3 + \beta_2)}, \quad g_{12} = -\frac{2\alpha_3\left[(\beta_2 - 3\alpha_3)\gamma_2 + 3\beta_2(5\alpha_3 + \beta_2)t_1\right]}{(\alpha_3 + \beta_2)(9\alpha_3 + \beta_2)}, \quad (C.4)$$

$$g_{13} = \frac{2\alpha_3 (3\alpha_3 - \beta_2)}{(\alpha_3 + \beta_2)(9\alpha_3 + \beta_2)}, \quad g_{22} = -\frac{2\beta_2^2 (5\alpha_3 + \beta_2)}{(\alpha_3 + \beta_2)(9\alpha_3 + \beta_2)}, \quad (C.5)$$

$$g_{14} = 9\, t_1\, g_{13}, \quad g_{23} = -\frac{2\left[\gamma_2(5\alpha_3 + \beta_2) + \beta_2(\beta_2 + 13\alpha_3)t_1\right]}{(\alpha_3 + \beta_2)(9\alpha_3 + \beta_2)}, \quad (C.6)$$

$$g_{24} = \frac{\beta_2}{\alpha_3} g_{13}, \quad g_{33} = \frac{g_{22}}{\beta_2^2}, \quad (C.7)$$

$$g_{34} = 3t_1 g_{33}, \quad g_{44} = \frac{g_{11}}{\alpha_3^2}. \quad (C.8)$$

Note that special relations among the coefficients can change the asymptotic behaviour. E.g., for all models in [141] we have

$$\beta_2 = 3\alpha_3. \quad (C.9)$$

yielding

$$g_{13} = g_{14} = g_{24} = 0. \quad (C.10)$$

Specifically, for the mirror quintic values, the expansion of $\mathcal{G}_t$ is

$$\mathcal{G}_t = \begin{pmatrix} \frac{5}{6} t_2^3 + \mathcal{O}(t_2) & \frac{5t_1}{2} t_2 + \mathcal{O}(t_2^{-1}) & -\left(\frac{5}{6} + t_1^2\right)\frac{1}{t_2} + \mathcal{O}(t_2^{-3}) & \frac{-10t_1^3 - 25t_1 + 12i\delta_3}{10}\frac{1}{t_2^3} + \mathcal{O}(e^{-t_2}) \\ \cdot & \frac{5}{2}t_2 + \mathcal{O}(t_2^{-1}) & -\frac{10t_1 + 11}{5}\frac{1}{t_2} + \mathcal{O}(t_2^{-3}) & \frac{-30t_1^2 - 66t_1 + 25}{10}\frac{1}{t_2^3} + \mathcal{O}(t_2^{-5}) \\ \cdot & \cdot & \frac{2}{5}\frac{1}{t_2} + \mathcal{O}(t_2^{-3}) & \frac{6t_1}{5}\frac{1}{t_2^3} + \mathcal{O}(t_2^{-5}) \\ \cdot & \cdot & \cdot & \frac{6}{5}\frac{1}{t_2^3} + \mathcal{O}(t_2^{-5}) \end{pmatrix}. \quad (C.11)$$

The Kähler covariant derivative of the period vector has the expansion

$$\begin{pmatrix} D_t \Pi_3 \\ D_t \Pi_2 \\ D_t \Pi_1 \\ D_t \Pi_0 \end{pmatrix} \sim \begin{pmatrix} A_3\, t_2^2 + B_3\, t_2 + C_3 + \ldots \\ B_2 t_2 + C_2 + \ldots \\ C_1 + \frac{D_1}{t_2} + \ldots \\ \frac{D_0}{t_2} + \frac{E_0}{t_2^2} + \ldots \end{pmatrix}. \quad (C.12)$$

where

$$A_3 = -\frac{3}{2}\alpha_3, \quad B_3 = -\frac{i\alpha_3 t_1(3\alpha_3 - 5\beta_2)}{2(\alpha_3 + \beta_2)}, \quad B_2 = \frac{i\beta_2}{2}, \quad C_2 = -\frac{4\alpha_3 \beta_2 t_1}{\alpha_3 + \beta_2} - \frac{\gamma_2}{2},$$

$$C_1 = -\frac{1}{2}, \quad D_1 = \frac{it_1(9\alpha_3 + \beta_2)}{2(\alpha_3 + \beta_2)}, \quad D_0 = \frac{3i}{2}, \quad E_0 = \frac{t_1(3\alpha_3 - \beta_2)}{\alpha_3 + \beta_2}. \quad (C.13)$$

Model	α_3	γ_3	δ_3	β_2	γ_2	δ_2
Mirror Quintic:	$-\frac{5}{6}$	$-\frac{25}{12}$	$\frac{200\zeta(3)}{(2\pi)^3}$	$-\frac{5}{2}$	$-\frac{11}{2}$	$\frac{25}{12}$
Model 12:	$-\frac{2}{3}$	$-\frac{5}{3}$	$\frac{18\zeta(3)}{(\pi)^3}$	-2	-5	$\frac{5}{3}$

Table C.1: Expansion coefficients around the LCS points for the considered one-parameter models.

C.1.2 Coefficients of metric $\mathcal{G}_z$ of the two-parameter model

The expansion of the metric of the complex structure moduli space of the model $\mathcal{M}_{(86,2)}$ near to the LCS is given in formula (6.64). Here we list its coefficients a_{ij}:

$$a_{11} = \frac{17}{6},$$
$$a_{12} = -\frac{545}{864}x_1 - \frac{61477}{10368}y_1,$$
$$a_{13} = -\frac{109}{72}x_1 - \frac{545}{864}y_1,$$
$$a_{14} = -\frac{109}{144}x_1 y_1 - \frac{545}{1728}y_1^2 - \frac{545}{1728},$$
$$a_{15} = \frac{545}{1728}x_1 y_1 - \frac{26651}{20736}y_1^2 - \frac{11963}{20736},$$
$$a_{16} = -\frac{109}{144}x_1 y_1^2 - \frac{109}{288}x_1 - \frac{545}{1728}y_1^3 - \frac{2725}{3456}y_1,$$
$$a_{22} = \frac{61477}{10368},$$
$$a_{23} = \frac{545}{864},$$
$$a_{24} = \frac{109}{144}x_1 + \frac{545}{864}y_1 + \frac{109}{1152},$$
$$a_{25} = -\frac{545}{1728}x_1 + \frac{26651}{10368}y_1 - \frac{545}{13824},$$
$$a_{26} = \frac{545}{576}y_1^2 + \frac{109}{72}x_1 y_1 + \frac{109}{576}y_1 - \frac{2725}{3456}, \quad a_{33} = \frac{109}{72},$$
$$a_{34} = \frac{109}{144}y_1,$$
$$a_{35} = -\frac{545}{1728}y_1,$$
$$a_{36} = \frac{109}{144}y_1^2 - \frac{109}{288},$$
$$a_{44} = \frac{109}{576},$$
$$a_{45} = -\frac{545}{6912},$$
$$a_{46} = \frac{109}{288}y_1,$$
$$a_{55} = \frac{61477}{82944},$$
$$a_{56} = \frac{109}{288}x_1,$$
$$a_{66} = \frac{109}{288}. \tag{C.14}$$

Bibliography

[1] M. Kreuzer and H. Skarke, "PALP: A package for analyzing lattice polytopes with applications to toric geometry," *Comput. Phys. Commun.*, vol. 157, pp. 87–106, 2004, math/0204356.

[2] A. P. Braun, J. Knapp, H. Skarke, and N.-O. Walliser, "PALP – a User Manual," 2012, 1205.4147.

[3] V. Balasubramanian, P. Berglund, J. P. Conlon, and F. Quevedo, "Systematics of moduli stabilisation in Calabi-Yau flux compactifications," *JHEP*, vol. 03, p. 007, 2005, hep-th/0502058.

[4] R. Blumenhagen, S. Moster, and E. Plauschinn, "Moduli Stabilisation versus Chirality for MSSM like Type IIB Orientifolds," *JHEP*, vol. 01, p. 058, 2008, 0711.3389.

[5] S. Ashok and M. R. Douglas, "Counting flux vacua," *JHEP*, vol. 0401, p. 060, 2004, hep-th/0307049.

[6] M. Grana, "Flux compactifications in string theory: a comprehensive review," *Phys.Rept.*, vol. 423, pp. 91–158, 2006, hep-th/0509003.

[7] M. R. Douglas and S. Kachru, "Flux compactification," *Rev. Mod. Phys.*, vol. 79, pp. 733–796, 2007, hep-th/0610102.

[8] R. Blumenhagen, B. Kors, D. Lüst, and S. Stieberger, "Four-dimensional String Compactifications with D-Branes, Orientifolds and Fluxes," *Phys. Rept.*, vol. 445, pp. 1–193, 2007, hep-th/0610327.

[9] P. Candelas, G. T. Horowitz, A. Strominger, and E. Witten, "Vacuum Configurations for Superstrings," *Nucl.Phys.*, vol. B258, pp. 46–74, 1985.

[10] J. M. Maldacena and C. Nunez, "Supergravity description of field theories on curved manifolds and a no go theorem," *Int.J.Mod.Phys.*, vol. A16, pp. 822–855, 2001, hep-th/0007018.

[11] S. B. Giddings, S. Kachru, and J. Polchinski, "Hierarchies from fluxes in string compactifications," *Phys.Rev.*, vol. D66, p. 106006, 2002, hep-th/0105097.

[12] F. Denef, "Les Houches Lectures on Constructing String Vacua," 2008, 0803.1194.

[13] T. Weigand, "Lectures on F-theory compactifications and model building," *Class.Quant.Grav.*, vol. 27, p. 214004, 2010, 1009.3497.

[14] A. Sen, "F theory and the Gimon-Polchinski orientifold," *Nucl.Phys.*, vol. B498, pp. 135–155, 1997, hep-th/9702061.

[15] G. T. Horowitz and A. Strominger, "Black strings and P-branes," *Nucl.Phys.*, vol. B360, pp. 197–209, 1991.

[16] B. R. Greene, A. D. Shapere, C. Vafa, and S.-T. Yau, "Stringy Cosmic Strings and Noncompact Calabi-Yau Manifolds," *Nucl.Phys.*, vol. B337, p. 1, 1990.

[17] A. Braun, A. Hebecker, and H. Triendl, "D7-Brane Motion from M-Theory Cycles and Obstructions in the Weak Coupling Limit," *Nucl.Phys.*, vol. B800, pp. 298–329, 2008, 0801.2163.

[18] T. W. Grimm, "The N=1 effective action of F-theory compactifications," *Nucl.Phys.*, vol. B845, pp. 48–92, 2011, 1008.4133.

[19] A. Sen, "Orientifold limit of F theory vacua," *Nucl.Phys.Proc.Suppl.*, vol. 68, pp. 92–98, 1998, hep-th/9709159.

[20] V. Danilov, "The geometry of toric varieties," *Russian Mathematical Surveys*, vol. 33, p. 97, 1978.

[21] W. Fulton, *Introduction to Toric Varieties*. No. 131 in Annals of mathematical studies, Princeton University Press, 1993.

[22] D. Cox, "Minicourse on Toric Varieties." http://www.amherst.edu/~dacox.

[23] D. Cox and S. Katz, *Mirror Symmetry and Algebraic Geometry*, vol. 68 of *Mathematical Surveys and Monographs*. Berlin/Heidelberg: American Mathematical Society, 1999.

[24] M. Kreuzer, "Toric geometry and Calabi-Yau compactifications," *Ukr.J.Phys.*, vol. 55, p. 613, 2010, hep-th/0612307.

[25] V. Bouchard, "Lectures on complex geometry, Calabi-Yau manifolds and toric geometry," 2007, hep-th/0702063. An older version of these notes was published in the Proceedings of the Modave Summer School in Mathematical Physics 2005.

[26] D. Huybrechts, *Complex geometry*. Springer, 2005.

[27] J. Treutlein, *Birationale Eigenschaften generischer Hyperfaechen in algebraischen Tori*. PhD thesis, Eberhard-Karls-Universitaet Tuebingen, 06 2010. http://tobias-lib.uni-tuebingen.de/volltexte/2010/4897/.

[28] V. Batyrev, "Dual Polyhedra and Mirror Symmetry for Calabi-Yau Hypersurfaces in Toric Varieties," *J. Algebraic Geom.*, vol. 3, pp. 493–535, 1994, math/0204356.

[29] M. Kreuzer and H. Skarke, "Classification of reflexive polyhedra in three-dimensions," *Adv.Theor.Math.Phys.*, vol. 2, pp. 847–864, 1998, hep-th/9805190.

[30] M. Kreuzer and H. Skarke, "Complete classification of reflexive polyhedra in four dimensions," *Adv. Theor. Math. Phys.*, vol. 4, pp. 1209–1230, 2002, hep-th/0002240.

[31] H. Skarke. Private communication.

[32] C. Wall, "Classification Problems in Differential Topology. V. On Certain 6-Manifolds.," *Inventiones Mathematicae*, vol. 1, p. 355, 1966.

[33] V. Batyrev and M. Kreuzer, "Constructing new Calabi-Yau 3-folds and their mirrors via conifold transitions," 2008, 0802.3376.

[34] "Calabi Yau data: Tools and data for (toric) Calabi-Yau varieties, Landau-Ginzburg models, and related objects." http://hep.itp.tuwien.ac.at/~kreuzer/CY/.

[35] A. P. Braun and N.-O. Walliser, "A new offspring of PALP," 2011, 1106.4529.

[36] T. Oda and H. Park, "Linear Gale transforms and Gel'fand-Kapranov-Zelevinskij decompositions," *Tohoku Mathematical Journal*, vol. 43, no. 3, pp. 375–399, 1991.

[37] P. Berglund, S. H. Katz, and A. Klemm, "Mirror symmetry and the moduli space for generic hypersurfaces in toric varieties," *Nucl.Phys.*, vol. B456, pp. 153–204, 1995, hep-th/9506091.

[38] W. Decker, G.-M. Greuel, G. Pfister, and H. Schönemann, "SINGULAR 3-1-5 — A Computer Algebra System for Polynomial Computations," 2012. http://www.singular.uni-kl.de.

[39] T. Oda, *Convex bodies and algebraic geometry: an introduction to the theory of toric varieties*. Springer-Verlag, 1988.

[40] L. Billera, P. Filliman, and B. Sturmfels, "Constructions and complexity of secondary polytopes," *Advances in Mathematics*, vol. 83, no. 2, pp. 155–179, 1990.

[41] I. Gel'fand, M. Kapranov, and A. Zelevinsky, *Discriminants, resultants, and multidimensional determinants*. Springer, 1994.

[42] J. Rambau, "TOPCOM: Triangulations of point configurations and oriented matroids," in *Mathematical Software—ICMS 2002* (A. M. Cohen, X.-S. Gao, and N. Takayama, eds.), pp. 330–340, World Scientific, 2002.

[43] W. Stein *et al.*, *Sage Mathematics Software (Version 4.7)*. The Sage Development Team, 2011. http://www.sagemath.org.

[44] F. Hirzebruch, *Topological methods in algebraic geometry*. Mathematical Surveys and Monographs, Springer-Verlag, 1995.

[45] Wolfram Research, Inc., "Mathematica edition: Version 8.0," http://www.wolfram.com/.

[46] J. Knapp, M. Kreuzer, C. Mayrhofer, and N.-O. Walliser, "Toric Construction of Global F-Theory GUTs," *JHEP*, vol. 1103, p. 138, 2011, 1101.4908.

[47] S. Kachru, R. Kallosh, A. Linde, and S. P. Trivedi, "De Sitter vacua in string theory," *Phys. Rev.*, vol. D68, p. 046005, 2003, hep-th/0301240.

[48] M. Wijnholt, "F-Theory, GUTs and Chiral Matter," 2008, 0809.3878.

[49] D. S. Freed and E. Witten, "Anomalies in string theory with D-branes," *Asian J.Math*, vol. 3, p. 819, 1999, hep-th/9907189.

[50] M. Cicoli, J. P. Conlon, and F. Quevedo, "General Analysis of LARGE Volume Scenarios with String Loop Moduli Stabilisation," *JHEP*, vol. 10, p. 105, 2008, 0805.1029.

[51] S. Gukov, C. Vafa, and E. Witten, "CFT's from Calabi-Yau four-folds," *Nucl. Phys.*, vol. B584, pp. 69–108, 2000, hep-th/9906070.

[52] K. Becker, M. Becker, M. Haack, and J. Louis, "Supersymmetry breaking and alpha'-corrections to flux induced potentials," *JHEP*, vol. 06, p. 060, 2002, hep-th/0204254.

[53] K. Bobkov, "Volume stabilization via alpha' corrections in type IIB theory with fluxes," *JHEP*, vol. 05, p. 010, 2005, hep-th/0412239.

[54] M. Cicoli, J. P. Conlon, and F. Quevedo, "Systematics of String Loop Corrections in Type IIB Calabi-Yau Flux Compactifications," *JHEP*, vol. 01, p. 052, 2008, 0708.1873.

[55] J. Evslin, "What does(n't) K-theory classify?," 2006, hep-th/0610328.

[56] R. Minasian and G. W. Moore, "K-theory and Ramond-Ramond charge," *JHEP*, vol. 11, p. 002, 1997, hep-th/9710230.

[57] F. Denef and G. W. Moore, "Split states, entropy enigmas, holes and halos," 2007, hep-th/0702146.

[58] N. Akerblom, R. Blumenhagen, D. Lüst, and M. Schmidt-Sommerfeld, "D-brane Instantons in 4D Supersymmetric String Vacua," *Fortsch. Phys.*, vol. 56, pp. 313–323, 2008, 0712.1793.

[59] R. Blumenhagen, M. Cvetic, and T. Weigand, "Spacetime instanton corrections in 4D string vacua - the seesaw mechanism for D-brane models," *Nucl. Phys.*, vol. B771, pp. 113–142, 2007, hep-th/0609191.

[60] N. Akerblom, "D-instantons and effective couplings in intersecting d-brane models," *Fortschritte der Physik*, vol. 56, no. 11-12, pp. 1065–1142, 2008.

[61] S. Kachru and D. Simic, "Stringy Instantons in IIB Brane Systems," 2008, 0803.2514.

[62] R. Argurio, M. Bertolini, S. Franco, and S. Kachru, "Metastable vacua and D-branes at the conifold," *JHEP*, vol. 06, p. 017, 2007, hep-th/0703236.

[63] R. Argurio, M. Bertolini, G. Ferretti, A. Lerda, and C. Petersson, "Stringy Instantons at Orbifold Singularities," *JHEP*, vol. 06, p. 067, 2007, 0704.0262.

[64] M. Bianchi, F. Fucito, and J. F. Morales, "D-brane Instantons on the T^6/Z_3 orientifold," *JHEP*, vol. 07, p. 038, 2007, 0704.0784.

[65] L. E. Ibanez, A. N. Schellekens, and A. M. Uranga, "Instanton Induced Neutrino Majorana Masses in CFT Orientifolds with MSSM-like spectra," *JHEP*, vol. 06, p. 011, 2007, 0704.1079.

[66] C. Petersson, "Superpotentials From Stringy Instantons Without Orientifolds," *JHEP*, vol. 05, p. 078, 2008, 0711.1837.

[67] P. Griffiths and J. Harris, *Principles of algebraic geometry*, vol. 1994. Wiley, 1978.

[68] I. Garcia-Etxebarria and A. M. Uranga, "Non-perturbative superpotentials across lines of marginal stability," *JHEP*, vol. 01, p. 033, 2008, 0711.1430.

[69] I. Garcia-Etxebarria, F. Marchesano, and A. M. Uranga, "Non-perturbative F-terms across lines of BPS stability," *JHEP*, vol. 07, p. 028, 2008, 0805.0713.

[70] C. Beasley and E. Witten, "New instanton effects in supersymmetric QCD," *JHEP*, vol. 01, p. 056, 2005, hep-th/0409149.

[71] C. Beasley and E. Witten, "New instanton effects in string theory," *JHEP*, vol. 02, p. 060, 2006, hep-th/0512039.

[72] A. Collinucci, F. Denef, and M. Esole, "D-brane Deconstructions in IIB Orientifolds," *JHEP*, vol. 0902, p. 005, 2009, 0805.1573.

[73] R. Blumenhagen, V. Braun, T. W. Grimm, and T. Weigand, "GUTs in Type IIB Orientifold Compactifications," *Nucl. Phys.*, vol. B815, pp. 1–94, 2009, 0811.2936.

[74] T. W. Grimm and A. Klemm, "U(1) Mediation of Flux Supersymmetry Breaking," *JHEP*, vol. 10, p. 077, 2008, 0805.3361.

[75] M. Cicoli, C. Mayrhofer, and R. Valandro, "Moduli Stabilisation for Chiral Global Models," 2011, 1110.3333.

[76] R. Donagi and M. Wijnholt, "Model Building with F-Theory," 2008, 0802.2969.

[77] C. Beasley, J. J. Heckman, and C. Vafa, "GUTs and Exceptional Branes in F-theory - I," *JHEP*, vol. 01, p. 058, 2009, 0802.3391.

[78] C. Beasley, J. J. Heckman, and C. Vafa, "GUTs and Exceptional Branes in F-theory - II: Experimental Predictions," *JHEP*, vol. 01, p. 059, 2009, 0806.0102.

[79] J. J. Heckman, "Particle Physics Implications of F-theory," 2010, 1001.0577.

[80] B. Andreas and G. Curio, "From Local to Global in F-Theory Model Building," *J.Geom.Phys.*, vol. 60, pp. 1089–1102, 2010, 0902.4143.

[81] R. Blumenhagen, T. W. Grimm, B. Jurke, and T. Weigand, "Global F-theory GUTs," *Nucl. Phys.*, vol. B829, pp. 325–369, 2010, 0908.1784.

[82] J. Marsano, N. Saulina, and S. Schafer-Nameki, "Compact F-theory GUTs with U(1) (PQ)," *JHEP*, vol. 1004, p. 095, 2010, 0912.0272.

[83] T. W. Grimm, S. Krause, and T. Weigand, "F-Theory GUT Vacua on Compact Calabi-Yau Fourfolds," *JHEP*, vol. 1007, p. 037, 2010, 0912.3524.

[84] R. Blumenhagen, A. Collinucci, and B. Jurke, "On Instanton Effects in F-theory," *JHEP*, vol. 08, p. 079, 2010, 1002.1894.

[85] M. Cvetic, I. Garcia-Etxebarria, and J. Halverson, "Global F-theory Models: Instantons and Gauge Dynamics," *JHEP*, vol. 1101, p. 073, 2011, 1003.5337.

[86] H. Hayashi, T. Kawano, Y. Tsuchiya, and T. Watari, "More on Dimension-4 Proton Decay Problem in F-theory – Spectral Surface, Discriminant Locus and Monodromy," *Nucl.Phys.*, vol. B840, pp. 304–348, 2010, 1004.3870.

[87] T. W. Grimm and T. Weigand, "On Abelian Gauge Symmetries and Proton Decay in Global F-theory GUTs," *Phys.Rev.*, vol. D82, p. 086009, 2010, 1006.0226.

[88] J. Marsano, N. Saulina, and S. Schafer-Nameki, "A Note on G-Fluxes for F-theory Model Building," *JHEP*, vol. 11, p. 088, 2010, 1006.0483.

[89] Y.-C. Chung, "On Global Flipped SU(5) GUTs in F-theory," *JHEP*, vol. 1103, p. 126, 2011, 1008.2506.

[90] M. Cvetic, I. Garcia-Etxebarria, and J. Halverson, "On the computation of non-perturbative effective potentials in the string theory landscape: IIB/F-theory perspective," 2010, 1009.5386.

[91] S. Cecotti, C. Cordova, J. J. Heckman, and C. Vafa, "T-Branes and Monodromy," *JHEP*, vol. 1107, p. 030, 2011, 1010.5780.

[92] J. Marsano, "Hypercharge Flux, Exotics, and Anomaly Cancellation in F-theory GUTs," *Phys.Rev.Lett.*, vol. 106, p. 081601, 2011, 1011.2212.

[93] A. Collinucci and R. Savelli, "On Flux Quantization in F-Theory," 2010, 1011.6388.

[94] C.-C. Chiou, A. E. Faraggi, R. Tatar, and W. Walters, "T-branes and Yukawa Couplings," *JHEP*, vol. 1105, p. 023, 2011, 1101.2455.

[95] C.-M. Chen, J. Knapp, M. Kreuzer, and C. Mayrhofer, "Global SO(10) F-theory GUTs," *JHEP*, vol. 1010, p. 057, 2010, 1005.5735.

[96] K. Kodaira, "On compact analytic surfaces: II," *Annals of Mathematics*, pp. 563–626, 1963.

[97] J. Tate, *Algorithm for determining the type of a singular fiber in an elliptic pencil*, vol. Modular functions of one variable IV of *Lecture Notes in Mathematics*. Berlin/Heidelberg: Springer, 1975.

[98] A. Klemm, M. Kreuzer, E. Riegler, and E. Scheidegger, "Topological string amplitudes, complete intersection Calabi-Yau spaces and threshold corrections," *JHEP*, vol. 05, p. 023, 2005, hep-th/0410018.

[99] C. Cordova, "Decoupling Gravity in F-Theory," 2009, 0910.2955.

[100] V. Braun, "Discrete Wilson Lines in F-Theory," 2010, 1010.2520.

[101] H. Hayashi, T. Kawano, Y. Tsuchiya, and T. Watari, "Flavor Structure in F-theory Compactifications," *JHEP*, vol. 1008, p. 036, 2010, 0910.2762.

[102] M. Cvetic and J. Halverson, "TASI Lectures: Particle Physics from Perturbative and Non- perturbative Effects in D-braneworlds," 2011, 1101.2907.

[103] V. V. Batyrev and L. A. Borisov, "Mirror duality and string-theoretic Hodge numbers," 1995, alg-geom/9509009.

[104] http://hep.itp.tuwien.ac.at/f-theory/.

[105] C.-M. Chen and Y.-C. Chung, "Flipped SU(5) GUTs from E_8 Singularities in F-theory," *JHEP*, vol. 1103, p. 049, 2011, 1005.5728.

[106] J. Marsano, N. Saulina, and S. Schafer-Nameki, "F-theory Compactifications for Supersymmetric GUTs," *JHEP*, vol. 08, p. 030, 2009, 0904.3932.

[107] C. Mayrhofer, *Compactifications of Type IIB String Theory and F-Theory Models by Means of Toric Geometry*. PhD thesis, Vienna University of Technology, 11 2010. http://aleph.ub.tuwien.ac.at.

[108] R. Donagi and M. Wijnholt, "Higgs Bundles and UV Completion in F-Theory," 2009, 0904.1218.

[109] M. Alim, M. Hecht, H. Jockers, P. Mayr, A. Mertens, *et al.*, "Hints for Off-Shell Mirror Symmetry in type II/F-theory Compactifications," *Nucl.Phys.*, vol. B841, pp. 303–338, 2010, 0909.1842.

[110] T. W. Grimm, T.-W. Ha, A. Klemm, and D. Klevers, "Computing Brane and Flux Superpotentials in F-theory Compactifications," *JHEP*, vol. 04, p. 015, 2010, 0909.2025.

[111] T. W. Grimm, T.-W. Ha, A. Klemm, and D. Klevers, "Five-Brane Superpotentials and Heterotic / F-theory Duality," *Nucl.Phys.*, vol. B838, pp. 458–491, 2010, 0912.3250.

[112] H. Jockers, P. Mayr, and J. Walcher, "On N=1 4d Effective Couplings for F-theory and Heterotic Vacua," *Adv.Theor.Math.Phys.*, vol. 14, pp. 1433–1514, 2010, 0912.3265.

[113] M. Alim, M. Hecht, H. Jockers, P. Mayr, A. Mertens, *et al.*, "Type II/F-theory Superpotentials with Several Deformations and N=1 Mirror Symmetry," *JHEP*, vol. 1106, p. 103, 2011, 1010.0977.

[114] T. W. Grimm, A. Klemm, and D. Klevers, "Five-Brane Superpotentials, Blow-Up Geometries and SU(3) Structure Manifolds," *JHEP*, vol. 1105, p. 113, 2011, 1011.6375.

[115] R. Blumenhagen, B. Jurke, T. Rahn, and H. Roschy, "Cohomology of Line Bundles: A Computational Algorithm," *J.Math.Phys.*, vol. 51, p. 103525, 2010, 1003.5217.

[116] L. Susskind, "The Anthropic landscape of string theory," 2003, hep-th/0302219.

[117] A. Vilenkin, "The Birth of Inflationary Universes," *Phys.Rev.*, vol. D27, p. 2848, 1983.

[118] A. D. Linde, "Eternally Existing Selfreproducing Chaotic Inflationary Universe," *Phys.Lett.*, vol. B175, pp. 395–400, 1986.

[119] A. D. Linde, "Eternal chaotic inflation," *Mod.Phys.Lett.*, vol. A1, p. 81, 1986.

[120] A. Aguirre, M. C. Johnson, and A. Shomer, "Towards observable signatures of other bubble universes," *Phys.Rev.*, vol. D76, p. 063509, 2007, 0704.3473.

[121] S. Chang, M. Kleban, and T. S. Levi, "When worlds collide," *JCAP*, vol. 0804, p. 034, 2008, 0712.2261.

[122] S. M. Feeney, M. C. Johnson, D. J. Mortlock, and H. V. Peiris, "First Observational Tests of Eternal Inflation," *Phys.Rev.Lett.*, vol. 107, p. 071301, 2011, 1012.1995.

[123] S. M. Feeney, M. C. Johnson, D. J. Mortlock, and H. V. Peiris, "First Observational Tests of Eternal Inflation: Analysis Methods and WMAP 7-Year Results," *Phys.Rev.*, vol. D84, p. 043507, 2011, 1012.3667.

[124] K. Freese and D. Spolyar, "Chain inflation: 'Bubble bubble toil and trouble'," *JCAP*, vol. 0507, p. 007, 2005, hep-ph/0412145.

[125] K. Freese, J. T. Liu, and D. Spolyar, "Chain inflation via rapid tunneling in the landscape," 2006, hep-th/0612056.

[126] D. Chialva and U. H. Danielsson, "Chain inflation revisited," *JCAP*, vol. 0810, p. 012, 2008, 0804.2846.

[127] D. Chialva and U. H. Danielsson, "Chain inflation and the imprint of fundamental physics in the CMBR," *JCAP*, vol. 0903, p. 007, 2009, 0809.2707.

[128] A. Ashoorioon, "Observing the structure of the landscape with the cmb experiments," *Journal of Cosmology and Astroparticle Physics*, vol. 2010, p. 002, 2010.

[129] S.-H. Henry Tye, "A New view of the cosmic landscape," 2006, hep-th/0611148.

[130] A. R. Brown and A. Dahlen, "Small Steps and Giant Leaps in the Landscape," *Phys.Rev.*, vol. D82, p. 083519, 2010, 1004.3994.

[131] M. C. Johnson and M. Larfors, "An Obstacle to populating the string theory landscape," *Phys.Rev.*, vol. D78, p. 123513, 2008, 0809.2604.

[132] A. R. Brown and A. Dahlen, "The Case of the Disappearing Instanton," 2011, 1106.0527.

[133] J. P. Conlon, F. Quevedo, and K. Suruliz, "Large-volume flux compactifications: Moduli spectrum and D3/D7 soft supersymmetry breaking," *JHEP*, vol. 08, p. 007, 2005, hep-th/0505076.

[134] T. R. Taylor and C. Vafa, "R R flux on Calabi-Yau and partial supersymmetry breaking," *Phys.Lett.*, vol. B474, pp. 130–137, 2000, hep-th/9912152.

[135] U. H. Danielsson, N. Johansson, and M. Larfors, "Stability of flux vacua in the presence of charged black holes," *JHEP*, vol. 0609, p. 069, 2006, hep-th/0605106.

[136] A. Ceresole, G. Dall'Agata, A. Giryavets, R. Kallosh, and A. D. Linde, "Domain walls, near-BPS bubbles, and probabilities in the landscape," *Phys.Rev.*, vol. D74, p. 086010, 2006, hep-th/0605266.

[137] U. H. Danielsson, N. Johansson, and M. Larfors, "The World next door: Results in landscape topography," *JHEP*, vol. 0703, p. 080, 2007, hep-th/0612222.

[138] D. Chialva, U. H. Danielsson, N. Johansson, M. Larfors, and M. Vonk, "Deforming, revolving and resolving - New paths in the string theory landscape," *JHEP*, vol. 0802, p. 016, 2008, 0710.0620.

[139] M. C. Johnson and M. Larfors, "Field dynamics and tunneling in a flux landscape," *Phys.Rev.*, vol. D78, p. 083534, 2008, 0805.3705.

[140] A. Aguirre, M. C. Johnson, and M. Larfors, "Runaway dilatonic domain walls," *Phys.Rev.*, vol. D81, p. 043527, 2010, 0911.4342.

[141] P. Ahlqvist, B. R. Greene, D. Kagan, E. A. Lim, S. Sarangi, *et al.*, "Conifolds and Tunneling in the String Landscape," *JHEP*, vol. 1103, p. 119, 2011, 1011.6588.

[142] F. Denef and M. R. Douglas, "Distributions of flux vacua," *JHEP*, vol. 0405, p. 072, 2004, hep-th/0404116.

[143] B. S. Acharya and M. R. Douglas, "A Finite landscape?," 2006, hep-th/0606212.

[144] T. Eguchi and Y. Tachikawa, "Distribution of flux vacua around singular points in Calabi-Yau moduli space," *JHEP*, vol. 0601, p. 100, 2006, hep-th/0510061.

[145] G. Torroba, "Finiteness of Flux Vacua from Geometric Transitions," *JHEP*, vol. 0702, p. 061, 2007, hep-th/0611002.

[146] R. Bousso and J. Polchinski, "Quantization of four form fluxes and dynamical neutralization of the cosmological constant," *JHEP*, vol. 0006, p. 006, 2000, hep-th/0004134.

[147] M. R. Douglas, "The Statistics of string / M theory vacua," *JHEP*, vol. 0305, p. 046, 2003, hep-th/0303194.

[148] R. Blumenhagen, F. Gmeiner, G. Honecker, D. Lüst, and T. Weigand, "The Statistics of supersymmetric D-brane models," *Nucl.Phys.*, vol. B713, pp. 83–135, 2005, hep-th/0411173.

[149] F. Gmeiner, R. Blumenhagen, G. Honecker, D. Lüst, and T. Weigand, "One in a billion: MSSM-like D-brane statistics," *JHEP*, vol. 0601, p. 004, 2006, hep-th/0510170.

[150] M. R. Douglas and W. Taylor, "The Landscape of intersecting brane models," *JHEP*, vol. 0701, p. 031, 2007, hep-th/0606109.

[151] P. S. Aspinwall and R. Kallosh, "Fixing all moduli for M-theory on K3xK3," *JHEP*, vol. 0510, p. 001, 2005, hep-th/0506014.

[152] F. Denef, "Supergravity flows and D-brane stability," *JHEP*, vol. 0008, p. 050, 2000, hep-th/0005049.

[153] K. Dasgupta, G. Rajesh, and S. Sethi, "M theory, orientifolds and G - flux," *JHEP*, vol. 9908, p. 023, 1999, hep-th/9908088.

[154] K. Becker and M. Becker, "Supersymmetry breaking, M theory and fluxes," *JHEP*, vol. 0107, p. 038, 2001, hep-th/0107044.

[155] M. R. Douglas, J. Shelton, and G. Torroba, "Warping and supersymmetry breaking," 2007, 0704.4001.

[156] M. R. Douglas and G. Torroba, "Kinetic terms in warped compactifications," *JHEP*, vol. 0905, p. 013, 2009, 0805.3700.

[157] P. K. Tripathy and S. P. Trivedi, "Compactification with flux on K3 and tori," *JHEP*, vol. 0303, p. 028, 2003, hep-th/0301139.

[158] A. P. Braun, A. Hebecker, C. Ludeling, and R. Valandro, "Fixing D7 Brane Positions by F-Theory Fluxes," *Nucl.Phys.*, vol. B815, pp. 256–287, 2009, 0811.2416.

[159] L. Andrianopoli, R. D'Auria, S. Ferrara, and M. A. Lledo, "4-D gauged supergravity analysis of type IIB vacua on K3 x T**2 / Z(2)," *JHEP*, vol. 0303, p. 044, 2003, hep-th/0302174.

[160] P. S. Aspinwall, "K3 surfaces and string duality," pp. 421–540, 1996, hep-th/9611137.

[161] W. Barth, K. Hulek, C. Peters, and A. Van de Ven, *Compact complex surfaces*. Springer-Verlag, Berlin, 2004.

[162] C. Borcea, "Diffeomorphisms of a k3 surface," *Mathematische Annalen*, vol. 275, no. 1, pp. 1–4, 1986.

[163] T. Matumoto, "On diffeomorphisms of a K3 surface," *Algebraic and Topological theories - to the memory of Dr. Takehiko Miyaka, edited by M. Nagata et al*, pp. 616–621.

[164] S. Donaldson, "Polynomial invariants for smooth four-manifolds," *Topology*, vol. 29, no. 3, pp. 257–315, 1990.

[165] A. P. Braun, R. Ebert, A. Hebecker, and R. Valandro, "Weierstrass meets Enriques," *JHEP*, vol. 1002, p. 077, 2010, 0907.2691.

[166] I. Brunner and K. Hori, "Orientifolds and mirror symmetry," *JHEP*, vol. 11, p. 005, 2004, hep-th/0303135.

[167] E. G. Gimon and J. Polchinski, "Consistency Conditions for Orientifolds and D-Manifolds," *Phys. Rev.*, vol. D54, pp. 1667–1676, 1996, hep-th/9601038.

i want morebooks!

Buy your books fast and straightforward online - at one of world's fastest growing online book stores! Environmentally sound due to Print-on-Demand technologies.

Buy your books online at

www.get-morebooks.com

Kaufen Sie Ihre Bücher schnell und unkompliziert online – auf einer der am schnellsten wachsenden Buchhandelsplattformen weltweit! Dank Print-On-Demand umwelt- und ressourcenschonend produziert.

Bücher schneller online kaufen

www.morebooks.de

VDM Verlagsservicegesellschaft mbH
Heinrich-Böcking-Str. 6-8
D - 66121 Saarbrücken

Telefon: +49 681 3720 174
Telefax: +49 681 3720 1749

info@vdm-vsg.de
www.vdm-vsg.de

Printed by Books on Demand GmbH, Norderstedt / Germany